Standards-Based Activities With Scoring Rubrics:

Middle & High School English

Volume 2

Performance-Based Projects

Jacqueline Glasgow, Editor

EYE ON EDUCATION
6 DEPOT WAY WEST, SUITE 106
LARCHMONT, NY 10538
(914) 833–0551
(914) 833–0761 fax
www.eyeoneducation.com

Library of Congress Cataloging-in-Publication Data

Standards-based activities with scoring rubrics : middle and high school English / [edited by] Jacqueline Glasgow.
p. cm.
Includes bibliographical references.
Contents: v. 1. Performance-based portfolios. — v. 2. Performance-based projects.
ISBN 1-930556-28-4 (v. 1) — ISBN 1-930556-29-2 (v. 2)
1. English language—United States—Examinations. 2. Academic achievement—United States—Evaluation. 3. English language—Study and teaching (Secondary)—Standards—United States. I. Glasgow, Jacqueline, 1941–

LB1631.5 .S78 2002
428'.0074—dc21

2001051221

10 9 8 7 6 5 4 3 2 1

Editorial and production services provided by
Richard H. Adin Freelance Editorial Services
52 Oakwood Blvd., Poughkeepsie, NY 12603-4112
(845-471-3566)

Also Available from EYE ON EDUCATION

A Collection of Performance Tasks & Rubrics: Middle School Mathematics
Charlotte Danielson

A Collection of Performance Tasks & Rubrics: High School Mathematics
Charlotte Danielson and Elizabeth Marquez

Assessment in Middle and High School Mathematics: A Teacher's Guide
Daniel J. Brahier

Teacher Leader
Thomas Poetter and Bernard Badiali

Constructivist Strategies: Meeting Standards and Engaging Adolescent Minds
Foote, Vermette, and Battaglia

Performance Standards and Authentic Learning
Allan A. Glatthorn

Performance Assessment and Standards-Based Curricula: The Achievement Cycle
Allan A. Glatthorn

Collaborative and Cooperative Learning: Applications and Assessments
Dawn M. Snodgrass and Mary M. Bevevino

Performance-based Learning and Assessment in Middle School Science
K. Michael Hibbard

Best Practices from America's Middle Schools
Charles R. Watson

The Interdisciplinary Curriculum
Arthur K. Ellis and Carol J. Stuen

CONTRIBUTORS

Jacqueline Glasgow, Ph.D.
Associate Professor of English Education
Ohio University, Athens

Ruth McClain, Teacher
Paint Valley High School
Chillicothe, Ohio

Tom Flynn, Ph.D.
Associate Professor of English
Ohio University, Eastern

Margie Bush, Teacher
Shawnee High School
Lima, Ohio

Nanci Bush, Teacher
Solon High School
Solon, Ohio

Susan Malaska, Teacher
Shelby High School
Shelby, Ohio

Janie Reinart
English/Language Arts Consultant
S. Russell, Ohio

Linda Rice, Teacher
Lakeview High School
Cortland, Ohio

Joyce Rowland, Teacher
Bristol High School
Bristolville, Ohio

Colleen Ruggieri, Teacher
Boardman High School
Youngstown, Ohio

Carolyn Suttles, Teacher
Bristol High School
Bristolville, Ohio

Acknowledgments

Many thanks to the Ohio Department of Education (ODE) for funding the project that has resulted in two volumes of best practice performance-based assessments for Middle School and High School English teachers. We are indebted to Dr. J. Daniel Good, Director of the Center for Curriculum and Assessment at ODE, and Dr. Kent Minor, Assistant Director of Professional Development and Licensure at ODE, for their confidence in the Ohio Council of Teachers of English to produce this work. We thank Bob Taft, Governor of Ohio, and Dr. Susan Tave Zelman, Ohio Superintendent of Public Instruction, for privileging the research and development of performance-based assessment practices.

This work could not have been completed without the strong support of the Executive Board of the Ohio Council of Teachers of English and Language Arts (OCTELA). OCTELA members submitted entries, gathered at weekend retreats, and volunteered their services in drafting, revising, and editing the manuscript as it progressed. Through team effort, our understanding of performance-based assessments has emerged into the two volumes in your hands today.

Many thanks to Bob Sickles and the staff at Eye on Education for their support in making these publications the best possible. I'd also like to thank the external reviewers Helen Dale, Joan Heiss, Pat Rooney, Myra Vinson, and Jennifer Watson.

Jacqueline Glasgow

Table of Contents

1

OVERVIEW

In this era of state proficiencies and high-stakes testing, the burdens of preparing students for standardized assessments has, at times, seemed to rob teachers of their willingness to "teach on the edge." This book presents a rich array of performance-based and standards-based activities with ready-made scoring tools designed to ensure accountability and celebrate the joy of learning simultaneously. By aligning standards, performance tasks (i.e., the activities found throughout this book), and assessments according to Glatthorn's Achievement Cycle (Figure 1.1), teachers can demonstrate that teaching on the edge, where joy, interaction, and innovation are the heart of the classroom, not only "measures up to," but can indeed surpass (and be a whole lot more fun than) "teaching to the test."

This is a book written by teachers for teachers. It is not prescriptive; rather, it is informational and allows the individual district and teacher the freedom to accept, reject, and modify any of the suggestions. In fact, we hope that teachers will see the many opportunities provided for innovative and creative teaching and assessment that recognizes the interrelationship of curriculum and assessment presented in this book. We have included three types of performance assessments: (1) a single piece of writing, (2) portfolio collections (literary and personal), and (3) multimedia assessments (dramatic, multimedia, oral, and museum). The purpose of this book is to enhance the quality of language learning in secondary English classrooms that is anchored in Glatthorn's Model of Authentic Learning, NCTE/IRA Standards for Language Arts Programs, and Falk's Using Standards and Assessment to Learn.

Authentic Learning Occurs in a Standards-Based Curriculum for English Language Arts

Glatthorn suggests that authentic learning is more likely to occur if the required curriculum is "derived from quality standards" (1999, p. 28). The standards published by the National Council of Teachers of English and the International Reading Association (NCTE/IRA) list 12 quality standards to guide curriculum for English Language Arts. The vision guiding these standards is that all students must have the opportunities and resources to develop the language skills they need to pursue life goals and to participate fully as informed, productive members of society. The standards provide a foundation for curriculum

Figure 1.1 Glatthorn's Model of Authentic Learning and Assessment

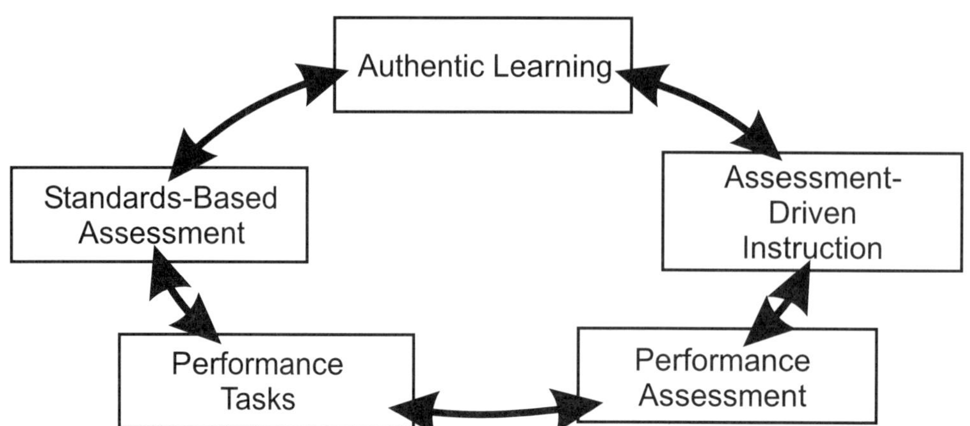

development. Standards for the English Language Arts (SELA) tells us that in the classroom we will find:

- ♦ students who are playing the roles of readers and writers, discovering how to shape their experience and to connect their experience to text;
- ♦ evidence of public audiences, classroom audiences, and personal audiences playing the roles of reader and responder to student work;
- ♦ subject matter, whether imaginary, public/civic, or academic and informational;
- ♦ different tools (computers, telephones, calculators, faxes) and editing groups;
- ♦ various texts both literary and nonliterary for reading, hearing, and viewing;
- ♦ language reference books on the structure of grammar (phonology, morphology, syntax) and test; and
- ♦ evidence of cognitive and metacognitive development in drafts from editing, discussion, and response groups, including learning logs, outlines, notes, and other forms (Myers and Spalding 1997, p. ix).

Literate people today must be effective communicators, critical thinkers, creative problem solvers, and lifelong learners. A curriculum built on these standards will encourage students to read a wide variety of print and nonprint texts, seek answers to meaningful questions, and appreciate the power and diversity of language as participating members of literacy communities. A curriculum built on these standards will enable students to develop their abilities in speaking, listening, reading, writing, viewing, and visual representation.

Authentic Learning Occurs if Based on a Meaningful Performance Task

According to Glatthorn, authentic learning is more likely to occur when students are engaged in a meaningful performance task. A performance task is "a complex open-ended problem posed for the student to solve as a means of demonstrating mastery" (Glatthorn 1999, p. 18). Marzano and Kendall (1996) identify the following characteristics of a performance task:

- requires knowledge to be applied to a specific situation
- provides necessary guidance and information to complete the task;
- specifies learning context (independent, pairs, small groups);
- specifies how students will demonstrate their findings or solution.

Once the performance tasks are identified, then the unit of instruction can be developed so that the students can do well when the assessment is made. Of course, the unit should be based on the approved curriculum guide for the school district. Performance tasks can be developed to prepare students for state performance assessments or to enrich the English Language Arts curriculum. Stiggins (1997) stresses that well-designed performance assessments are a highly effective teaching tool, significantly fostering student learning.

Authentic Learning Occurs Through Authentic Assessment

If we regard learning as a process of actively constructing meaning, rather than an accumulation of skills, then assessment tools must reflect our beliefs. Although traditional assessments, such as asking students to respond to writing prompts, are useful for preparing them to take proficiency tests, we encourage teachers to move toward more authentic assessment and evaluation. We want students to think, question, research, write, and share the meaning they have constructed.

Authentic assessment is geared toward methods that correspond as closely as possible to real-world experience. It was originally developed in the arts and apprenticeship systems, where assessment has always been based on performance. The instructor observes the student in the process of working on something real, provides feedback, monitors the student's use of feedback, and adjusts instruction and evaluation accordingly. Authentic assessment takes this principle of evaluating real work into all areas of the curriculum.

The rubric is one authentic assessment tool that is designed to simulate real-life activity where students are engaged in solving real-life problems. It is a formative type of assessment because the assessment rubric is available to students at the beginning of instruction serving as a learning tool. As students work toward meeting the expectations specified in the rubric, it becomes an ongoing part of the whole teaching and learning process. Students themselves can be involved in the assessment process using the rubric for both peer and self-assessment. As students become familiar with rubrics, they can assist in the rubric design process. This involvement empowers the students and, as a result, their learning becomes more focused and self-directed. Authentic assessment, therefore, blurs the lines between teaching, learning, and assessment. The advantages of using rubrics in assessment are that they:

- allow assessment to be more objective and consistent
- focus the teacher to clarify her/his criteria in specific terms
- clearly show the student how their work will be evaluated and what is expected
- promote student awareness of the criteria to use in assessing peer performance
- provide useful feedback regarding the effectiveness of the instruction
- provide benchmarks against which to measure and document progress

Rubrics can be created in a variety of forms and levels of complexity. However, they all contain common features that:

- focus on measuring a goal, standard, or benchmark (performance, behavior, or quality),
- provide a list of criteria for the project (assignment), and
- contain specific performance characteristics arranged in levels indicating the degree to which a standard has been met.

According to Starr (2000), a good rubric should:

- address all relevant content and performance objectives;
- define standards and help students achieve them by providing criteria with which they can evaluate their own work;
- be easy to understand and use;
- be applicable to a variety of tasks;
- provide all students with an opportunity to succeed at some level;
- yield consistent results, even when administered by different scorers.

The main drawback to using rubrics for assessment is that students want rubrics for everything they learn! (Adapted from Rubrics for Web Lessons: http://edweb.sdsu.edu/webquest/rubrics/welessons.htm.)

Standards-Based Assessment Activities and Scoring Rubrics

Based on the above theory and research, these two volumes contain a variety of standards-based and performance-based assessments with scoring rubrics. Also included are excellent models of student work to stimulate the imagination and demonstrate creative responses to the assessments. See Volume 1 for Traditional and Portfolio Assessments and Volume 2 for Performance-Based Project Assessments.

Traditional Assessment: Essays and Other Single Pieces of Writing (Volume 1)

Students read and write for specific purposes and to learn basic writing formats and conventions. Expository writing provides the best opportunity to directly teach the most important foundations of good writing: the development of coherence, logic, and ideas within the structural confines of the paragraph and the essay while supporting with evidence those claims that writers wish to make. If the essay is written as a one-shot event, the product is valued over the writing process. If the essay has traveled through peer editing and multiple drafting, the writing process has been observed and revision has been valued. For the teacher, this assessment provides a quick reference for remediation and intervention. In volume one of this book, prompts, scoring rubrics, and student samples are provided to help students meet the standards for the narrative, expository, and persuasive modes of writing found on many proficiency exams.

Personal and Literary Portfolio Collections (Volume 1)

These assessments consist of a collection of work, from writing folders to scrapbooks to mandated collections of work that showcase the student as a learner. Departing from the traditional writing portfolio that results in key artifacts from an entire year to show growth and development, these various types of portfolios fulfill a variety of purposes. Most of the portfolios in this book represent thematic units and therefore document the student's literacy by integrating reading, writing, art, music, technology, and other content areas. Portfolio assessments not only show the students' critical judgments and collaborative processes, but they also provide documentation of meeting district, state, and national standards. In volume one of this book, there are ideas for portfolios comprised of responses to literary collections such as cyber buddy journals, poetry notebooks, literature circles, and scrapbooks. The personal portfolio collections include an archaeological dig, "I-Search Paper," sketch books and other types of journals.

Performance-Based Project Assessments (Volume 2)

These performance-based project assessments encourage teachers to use the most progressive means possible for assessing a student's literacy performance. The activities included in this assessment ask students to demonstrate literacy through performance and presentations (mimes, dance); visual arts (music, photographs, sketches, cartoons); media literacy; multimedia performances; and technology projects in HyperStudio, PowerPoint, and/or Web pages.

These projects fulfill the performance objectives for composition, listening/visual literacy, oral communication, and reading mandated in the NCTE/IRA Standards. Although this type of assessment is not part of the Ohio High School Graduation Qualifying Exam, teachers and students who work with these kinds of performance activities are engaged in active responses that recognize and develop individual talents, thinking strategies, and modes of expression. Volume two of this book, provides examples of performance assessments through drama, multimedia projects, oral performances, and museums, as well as appropriate assessment rubrics.

Bibliography

Falk, B. (2000). *The heart of the matter: Using standards and assessment to learn.* Portsmouth, NH: Heinemann.

Farell, E. (1996). *In Standards for the English language arts.* Urbana, IL: National Council of Teachers of English/International Reading Association.

Glatthorn, A. A. (1999). *Performance standards authentic learning.* Larchmont, NY: Eye on Education.

Marzano, R.J., & Kendall, J.S. (1996). *A comprehensive guide to designing standards-based districts, schools, and classrooms.* Alexandria, VA: Association for Supervision and Curriculum Development.

Myers, M., & Spalding, E. (1997). *Standards exemplar series: Assessing student performance grades 9-12.* Urbana, IL: National Council of Teachers of English.

Rubrics for Web lessons [Online]. Available: http://edweb.sdsu.edu/webquest/rubrics/welessons.htm.

Standards for the English language arts (1996). Urbana, IL: National Council of Teachers of English/International Reading Association.

Starr, L. (2000). *A good rubric.* Education World.

Stiggins, R. J. (1997). *Student-centered classroom assessment (2nd ed.).* Columbus, OH: Merrill.

Part I

Dramatic Assessments

2

MEET-THE-AUTHORS CONVENTION

Jacqueline Glasgow

A journey of a thousand miles must begin with a single step.
Chinese Proverb

Overview of Authentic Learning

In this assessment, every student assumes the role of an author for the day. The author could be from American, British, or world literature. In this case, students will select or draw names of authors of Young Adult literature and keep their identity a secret. They must not disclose the name to other students, so that no one knows which student has chosen which author. They are expected to research and write a paper about the author's life, major works, and impact of the works on young adults. On the designated day, students show up at a Meet-the-Authors convention, and, in character, they attend a social gathering that gives them a chance to gather clues in determining which author is portrayed by which classmate. Then, in a general session, students give a brief presentation. At the end, there is a quiz to see who has identified the most authors at the conference. Give awards for the highest quiz score, best presentation, best costume (if costumes are part of the day), and best author profile paper.

NCTE/IRA Standards

In this assessment, students will:

- research a profile of an author and analyze her/his major works. In doing so, students will conduct research on issues and interests by generating ideas and questions, and by posing problems. They will gather, evaluate, and synthesize data from a variety of sources to communicate their discoveries in ways that suit their purpose and audience (NCTE/IRA Standard #7); students will use a variety of technological and informational resources to gather and synthesize information and to create and communicate knowledge (NCTE/IRA Standard #8); students will read a wide range of print and nonprint texts to build an understanding of texts, of themselves, and of the cultures of the U.S. and the world (NCTE/IRA Standard #1); students will read a wide range of literature from many periods in many genres to build an understanding of the many dimensions of human experience (NCTE/IRA Standard #2); students will apply a wide range of strategies to comprehend, interpret, evaluate, and appreciate texts (NCTE/IRA Standard #3); and, students will employ a wide range of strategies as they write and use different writing process elements appropriately to communicate with different audiences for a variety of purposes (NCTE/IRA Standard #5).

- prepare a book talk for the convention. In doing so, students will adjust their use of spoken, written, and visual language to communicate effectively with a variety of audiences and for different purposes (NCTE/IRA Standard #4); students will apply knowledge of language structure, language conventions, media techniques, figurative language, and genre to create, critique, and discuss print and nonprint texts (NCTE/IRA Standard #6), and students will use spoken, written, and visual language to accomplish their own purposes (NCTE/IRA Standard #12).
- identify authors represented by their classmates from oral and visual clues. In doing so, students develop an understanding of, and respect for, diversity in language use, patterns, and dialects across cultures, ethnic groups, geographic regions, and social roles (NCTE/IRA Standard #9), and students will participate as knowledgeable, reflective, creative, and critical members of a variety of literacy communities (NCTE/IRA Standard #11).

Performance Tasks

Performance Tasks for Profile of an Author and Analysis of Major Works

The purpose of this assignment is to prepare a profile of a critically acclaimed author and an analysis of some of the books this author has written. Choose a writer whose works appeal particularly to adolescents. This assessment was adapted from Anthony Manna, Kent State University, Kent, Ohio.

- Ask students to select or draw names of an author of any type of Young Adult literature (not only realistic novels). Students need to consult at least five sources for information on this author (poet, playwright, fantasists, short story writer, nonfiction writer, etc.). This part of the report will contain a profile of the author with attention to pertinent facts about her/his life, the reasons why she/he is a writer, and her/his goals as a writer. This part of the report is more than a mere copy of the information found in the sources they consulted. Students need to interpret what they read, looking for themes, topics, and issues regarding the person's life as a writer. Think of this part of the report as a character study. Students should document the information they take directly from their sources and include them on the "Works Cited" page. The length of this section is three to five typed, double-spaced pages. For suggested authors, see *List of Young Adult Authors.*

List of Young Adult Authors

Avi
Joan Bauer
Francesca Lia Block
Bruce Brooks
Sue Ellen Bridgers
Sook Nyul Choi
Brock Cole
Robert Cormier
Sharon Creech
Chris Crutcher
Christopher Paul Curtis
Sharon Draper
Paul Fleischman
Nancy Garden
Virginia Hamilton
Karen Hesse
S. E. Hinton
Will Hobbs
Paul Janeczko
M. E. Kerr
Lain Lawrence
Ellen Levine
Robert Lipsyte
Lois Lowry
Chris Lynch
Margaret Mahy
John Marsden
Carol Matas
Harry Mazer
Norma Fox Mazer
Robin McKinley
Walter Dean Myers
Lensey Namioka
Katherine Patterson
Gary Paulsen
Richard Peck
Phillip Pullman
Cynthia Rylant
Louis Sachar
Suzanne Fisher Staples
Gary Soto
Joyce Carol Thomas
Yoshiko Uchida
Cynthia Voigt
Virginia Euwer Wolff
Jacqueline Woodson
Jane Yolen
Paul Zindel

- In the second part, ask students to select three to five of the author's best works. Rather than choose these at random, the works should be the selections that provide a clear picture of the writer's skill, topics, major themes, issues, etc. Students could approach this part from a developmental perspective. This means examining the development of the writer's work over time. This part of the report will profile the writer's skill (with reference to specific elements of literature), the topics she/he best writes on, the major themes that emerge in the works they read, the issues she/he addresses, etc. The length of this section is three to five typed, double-spaced pages.
- The report will consist of a title page (student's name and a title that best captures the main theme of the report), the profile of the author,

the profile of the author's works, and a "Works Cited" page (done correctly in APA or MLA style) that includes the references used and the books written by the subject author. See *Assessment Rubric for Author Profile and Analysis of Major Works* on page 13.

Performance Tasks for Meeting-the-Authors Conference

After finishing the author profile paper, students begin preparation for the Meeting-the-Authors Conference. For this oral assessment, students role-play an author without giving away the author's identity, as well as try to identify the other authors attending the conference. Assess students on their oral presentation in the General Session, as well as their ability to interview and identify the authors at the Social Hour.

- To prepare for the Social Hour, students should agree on common interview questions for which they are accountable for the answers for their own author, as well as have questions to ask other authors. Interview questions might include: Where did you grow up? Did you like school? What were your favorite books? Did you play sports? What were you like as a kid? Where do you get your ideas? Where do you live now? What are your hobbies? What is your greatest accomplishment? What books have you written? What genres of books have you written? Who are your main characters? Any question goes except for: "What is your name?" Students are not to disclose their identities. Because most authors have Web sites giving such personal information, access to the Internet is the best resource.
- To prepare for the General Session, students should agree on the format for their presentations. They introduce themselves by giving some personal information about the author's life. They do a booktalk about one of their major works. Students could be required to wear a costume that would give clues as to their identity. Or, they could come dressed as one of the author's main characters. Since most of the Young Adult authors are still living in this modern age, requiring students to bring an artifact from one of the author's major works seems more appropriate. The artifact has to appear in the book they are talking about and be a significant clue; it cannot be inferred or metaphorical. For instance, for Cormier's Chocolate War, students could bring a box of chocolates or a more subtle artifact such as the Bible to represent the setting in the Catholic High School. They could dress as a Catholic monk, if costumes are to be part of the assessment. Decide the length of the presentation and the length of the question and answer period. See *Guidelines for a Booktalk* on page 14 and Becky Reid's written *Booktalk on Z for Zachariah* (by O'Brien) on page 14.

Assessment Rubric for Author Profile and Analysis of Major Works

	Exemplary (4)	*Accomplished (3)*	*Developing (2)*	*Beginning (1)*	*Score*
Author Profile	Profiles the author's life, interests, goals as a writer. Interpretative discussion of the writer's themes, interests, topics, genres, etc. Impressive research and documentation of sources.	Adequate information about the author's life, interests, and goals as a writer. Good analysis of the writer's themes, interests, topics, genres, etc. Adequate research and documentation of credible resources.	Reads more like a report than a character study. Important information is missing, even though there is a general discussion of the author's works. Contains questionable research sources.	Includes little essential information. No evidence that the writer has read about the author. No evidence of research.	__ x 8 = __ out of 32
Profile of Works	Profiles the writer's craft, topics, major themes, and issues emerging from the books read. Insight into key issues in the author's development as a writer. Provides textual evidence for key points.	Adequate discussion of the writer's craft, themes, and issues. Insights are keen, though not outstanding, concerning the author's development as a writer. Provides limited textual evidence for key points.	Important elements are missing from this analysis. Discussion based on common knowledge about the author's works. No textual evidence for key points.	Not obvious that the writer has read the works. Limited discussion of author's works. No reference to the works.	__ x 10 __ out of 40
Works Cited	Impressive citations of research from reliable sources. Correct MLA or APA format.	Cited at least 5 scholarly sources. Correct MLA or APA format.	Sources are questionable and/or incorrectly documented.	No sources cited.	__ x 4 = __ out of 16
Conventions	Correct sentence structure, grammar, punctuation, spelling.	Mostly correct sentence structure, grammar, punctuation, spelling.	Frequent errors in sentence structure, grammar, punctuation, spelling.	Many errors in sentence structure, grammar, punctuation, spelling.	__ x 3 = __ out of 12
			Total Points (100 possible)		
			Grade Scale	A = B = C = D =	Grade:

Guidelines for a Booktalk

The purpose of a booktalk is an introduction to a book that entices others to read it; its goal is to create interest, not to tell stories. A booktalk is not a book review, a book report, or a book analysis. It does not judge the book's merits; it does not reveal the ending. When making a booktalk, the assumption is that you like the book; otherwise how can you expect to "sell" it to others? A booktalk should last three to five minutes.

- Once you have read a book, think of a creative way to hook the potential reader.
- Select for oral reading a short passage from the text that gives a sample of the author's writing tone and style and that invites the listener to want to know more.
- Prepare to share an artifact that actually appears in the book, not an inference or metaphor.
- Either memorize your talk or put key points on note cards. Practice before you present.
- If a written report is required, include the following: bibliographic documentation for the book; brief summary that you may or may not use in your talk, but that includes all the significant details; and written rendition of the booktalk itself.

Booktalk on *Z for Zachariah*

What would you do if you were the only survivor of a nuclear holocaust? Sixteen-year-old Ann is sure she is the only human being left on earth after a weeklong nuclear war. Her family is dead. For a year she has managed to efficiently survive on her own. The hidden valley in which she lives has miraculously survived the deadly radiation. However, one day Ann notices smoke in the sky and realizes that someone-a survivor-is out there. As a precautionary, she leaves her house appearing as if it were unlived in. Supplied with food from the nearby small store, Ann sits and watches.

A man dressed in a green body suit and carrying a wagon arrives and enters her house. He sets up a tent and begins to bathe in a contaminated stream. As Ann secretly watches all this, Faro, the family dog, surprises her by suddenly appearing. The man retreats into the tent he has set up and does not even venture out. He is obviously very ill. Should she take care of him?

Ann, armed with her gun, goes down to see him. She decides to talk to the man who later identifies himself as John Loomis. The stream has given him a high dose of radiation. Being a radiation expert from Cornell University, he informs her that he will become anemic and very susceptible to germs. He will only get worse before he gets better. Questioning him, Ann discovers that Loomis worked for a Nobel Prize winning scientist. As a graduate student, Loomis worked on research in the area of plastics and polymers. Together they discovered a concept called "polarizing" to stop radiation. They developed one polarizing suit—the one Loomis has with him—before the war broke out.

Ann nurses the very ill Loomis. Loomis is delirious for days on end and spends the majority of his days sleeping. She increases the size of the garden, cooks for him, and even fantasizes about marrying him. With the help of some farming books, they get the tractor up and running.

One day, Ann discovers Loomis wildly shooting at the house, insisting that he hears an intruder and even aims the gun at her. He begins rambling on about a man named Edward. Ann gets a cold chill as she hears and connects the implication of his words. Loomis is a murderer. He shot his coworker, Edward. Edward had wanted to venture outdoors wearing the suit to check and see the fate of his family after the radiation.

Loomis is beginning to recover. Ann once again dreams of the possibility of becoming an English teacher and beginning a family. She bakes a cake in celebration of her birthday and Loomis's progress. He is beginning to walk again. He is worried about the garden and begins to consider the valley his. However, Loomis is acting strangely. He lashes out when Ann asks to borrow the suit and begins acting very possessively of her. He is starting to control her. Loomis tries to attack her and she just barely escapes. Ann moves once again to her cave out of fear of this man. She continues to supply him with food, but she is constantly on the alert. Loomis takes advantage of the dog to try to track her down. Bit by bit, Loomis reveals his madness. What should she do? Should she steal the green suit and leave the valley? Should she stay and try to reconcile with this man? Does she have a moral obligation to repopulate the earth? Should she shoot him and continue to live peacefully by herself? You'll have to read the book to find out.

- To prepare for the quiz contest, students should be given a class roster on a sheet of paper that is folded in half lengthwise to record author identification. They should carry this card around at the Social Hour and write down the authors their classmates are portraying as they interview, ask questions, and identify the authors. They continue trying to guess the authors throughout the presentations at the general session. At the end of the day, the quizzes are collected and graded. The person with the highest score receives a prize. Prizes can also be given for best costume, best artifact, best presentation, and best research paper. See *Assessment Rubric for the General Session Booktalk.*

Bibliography

Manna, A. (1992). *Young adult literature course materials.* Kent, OH: Kent State University.

Nilsen, A. P., & Donelson, K. L. (2001). *Literature for today's young adults,* (6th ed.). New York: Longman.

O'Brien, R. C. (1974). *Z for Zachariah.* New York: Aladdin.

Assessment Rubric for the General Session Booktalk

	Exemplary (4)	*Proficient (3)*	*Apprentice (2)*	*Novice (1)*	*Score*
Organization	Presentation began with a hook; logical progression of ideas; read an effective passage; gave a hint of the plot; did not reveal the ending	Presentation includes an introduction followed by a relatively organized progression of ideas; read an appropriate passage, didn't reveal the ending	Presenter "gushed" over the book; was not carefully prepared; tried to ad lib the booktalk; told too much of the story; revealed the ending	The message is so disorganized you cannot understand most of the message	__ x 5 = (20 points)
Content	Demonstrated a thorough understanding of the book; stayed in character giving creative clues of the author's identity	Demonstrated a good understanding of the book; stayed in character giving appropriate clues of the author's identity	Demonstrated a lack of understanding of the book; stayed in character, but gave ambiguous or misleading clues of the author's identity	Not obvious that the presenter has read the book; not enough content; speaker says practically nothing	__ x 5 = (20 points)
Artifact	Shared a creative artifact for the book	Shared an adequate artifact for the book	Failed to share an artifact or shared an obscure artifact	No artifact available	__ x 5 = (20 points)
Eye Contact	Student establishes good eye contact with audience and makes audience members feel as if they are being spoken to	During most of the presentation, student maintains eye contact with audience	Student establishes minimum eye contact with audience	No eye contact; read a paper, or avoided eye contact	__ x 2 = (8 points)
Vocal Projection	Student enunciates clearly, uses expression, and is clearly heard by the audience; speaker uses delivery to enhance the meaning of the message	Student speaks clearly, but with little expression; volume is not too low or too loud	Student does not enunciate and is difficult to hear; the rate is too fast or too slow	The student exhibits many disfluencies, such as "ahs," "uhms," or "you knows"	__ x 2 = (8 points)
Body Language	Student uses gestures and mannerisms to enhance the delivery and does so naturally and effectively	Student uses some gestures and mannerisms to enhance the delivery	Student uses few or inappropriate mannerisms that detract from the presentation	Student exhibits inappropriate behavior	__ x 2 = (8 points)
Costume of Main Character	Creative costume that represents an accurate portrayal of the character	Costume shows some accurate aspects of the character	Costume misrepresents or is not relevant to revealing the character	No costume; appeared in regular, school clothes	__ x 4 = (16 points)

Total Points (100 possible)

Grade Scale
A=
B=
C=
D=

3

INQUIRY INTO "THE DUST BOWL" PROCESS DRAMA

Jacqueline Glasgow

It has long come to my attention that people of accomplishment rarely sat back and let things happen to them. They went out and happened to things.

Elinor Smith

Adapted from NCTE presentation by Brian Edmiston and Patricia Enciso; Course Materials (Criticism of Children's Literature) by Anthony Manna; and extended and adapted by Jacqueline Glasgow

Overview of Authentic Learning

The fictional context for this drama requires students to assume the role of photojournalists in 1936 whose work has been to document people's struggles during "the Depression." They have been invited by Eleanor Roosevelt to join a Presidential Task Force to advise the President about the details of a bill to be presented in Congress to address the causes and consequences of "The Dust Bowl." Materials for the role-play come from: *Out of the Dust* (1997), by Karen Hesse; *Children of the Dust Bowl* (1992), by Jerry Stanley, *America at the Cross Roads; Great Photographs from the Thirties,* by J. Prescott, and *Eleanor Roosevelt: A Life of Discovery,* by Russell Freedman. The teacher prepares the materials and leads the class through the process drama as inquiry tasks. According to Cecily O'Neill, "process drama permits direct encounter with the event, a range of role-taking, and an encounter with the power of drama without necessarily demanding the immediate display of sophisticated acting techniques" (*Drama Worlds,* Heinemann, 1995, p. xiv). The emphasis on process rather than product results in creative dramatics and improvisation. The directions for the role-play follow the materials list.

NCTE/IRA Standards

In this assessment students will:

- role-play photojournalists of the Dust Bowl era. In doing so, students will apply a wide range of strategies to comprehend, interpret, evaluate, and appreciate texts. They will draw on their prior experience, their interactions with other readers and writers, their knowledge of word meaning and of other texts (NCTE/IRA Standard #3); students will adjust their use of spoken, written, and visual language to communicate effectively with a variety of audiences and for different purposes (NCTE/IRA Standard #4); students will apply knowledge of language structure, language, media techniques, figurative language, and genre to create, critique, and discuss print and nonprint texts (NCTE/IRA Standard #6); students will develop an understanding of, and respect for, diversity in language use, patterns, and dialects across cultures, ethnic groups, geographic regions, and social roles (NCTE/IRA Standard #9); students will participate as knowledgeable, reflective, creative, and critical members of a variety of literacy communities (NCTE/IRA Standard #11); and, students will use spoken, written, and visual language to accomplish their own purposes (NCTE/IRA Standard #12).
- read and interpret selections of literature set in the Dust Bowl. In doing so, students will read a wide range of print and nonprint texts to build an understanding of texts, of themselves, and of the cultures of the U.S. and the world (NCTE/IRA Standard #1); students will read a wide range of literature from many periods in many genres to build an understanding of the many dimensions of human experience (NCTE/IRA Standard #2); and, students will apply a wide range of strategies to comprehend, interpret, evaluate, and appreciate texts. They draw on their prior experience, their interactions with other readers and writers, their knowledge of word meaning and of other texts, their word identification strategies, and their understanding of textual features (NCTE/IRA Standard #3).
- write poetry and letters of recommendation. In doing so, students will adjust their use of spoken, written, and visual language to communicate effectively with a variety of audiences and for different purposes (NCTE/IRA Standard #4); students will employ a wide range of strategies as they write and use different writing process elements appropriately to communicate with different audiences for a variety of purposes (NCTE/IRA Standard #5); and, students will apply knowledge of language structure, language conventions, media techniques, figurative language, and genre to create, critique, and discuss print and nonprint texts (NCTE/IRA Standard #6).

Materials in the Room

- Photographs taken during the Depression mounted on a display board. These can be taken from www.snap.com (pictures of the Dust Bowl) or from Raven's, Rylant's or Prescott's books.
- Display a map of the Dust Bowl like the one found on page 5 of Stanley's book.
- Copies of a letter as if from Eleanor Roosevelt to photojournalists.
- Information about the Dust Bowl displayed on charts.
- Extracts from Children of the Dust Bowl on cards.
- Poems from Out of the Dust mounted on construction paper.

Planned Performance Tasks (Scenarios or Episodes)

Personal Responses to Photographs and Map from Dust Bowl

- "If you had taken these photos in the 1930's, what would you have wanted people to know, think, and feel about the Dust Bowl?"
- "How have you shown people's strength and dignity in the face of suffering and injustice?"
- "Who was affected most by the Dust Bowl?"
- Read the following quote (from blackboard or overhead transparency):

  "I felt I was becoming a slave to the land. But I held on to the thought that this land had to be stopped from blowing. Often I was so full of dust that I drove blind, unable to see even the radiator cap on my tractor or hear the roar of the engines. But I kept driving on and on, by guess and instinct. I was making my last stand in the Dust Bowl."
- "If you had been part of one of these farm families during the '30s, do you think you would have wanted to stay on your farm or leave? Why or why not? What would you lose by leaving? What would you gain?"

Agreement to Imagine Together

"Imagine that you have just received this letter from Eleanor Roosevelt (see *Letter from Eleanor Roosevelt to the Photojournalists* on page 24.) Consider:"

- "Why are you going to the meeting? What is your plan?"
- "How do you think we should document the consequences and causes of peoples' suffering from the Dust Bowl?"

- "How do you avoid taking advantage of people?"
- "What principles do you think should be followed in proposing legislation?"

Gathering Data: Pairs Reading and Response in Tableaux

In pairs, students each read a different extract from Stanley's Children of the Dust Bowl.

- "Imagine that these are the results of your investigations as you have spent some time taking photographs. Using your bodies to represent the people in the photos, create two pictures (tableaux) taken some time apart in the same place where you tried to show the same people."
- Give students 5 to 7 minutes to plan and rehearse their tableaux.

Interpretation of Images

- "As photojournalists, tell us what happened, what you saw, and what you overheard."
- Each pair performs their tableaux moving from the first frame to the second. The students interpret their own, and others', images. (Audience could write captions for each set of tableaux. Teacher could tap the shoulder of one member of the tableau and allow that person to speak in role: What are her/his thoughts? What emotions are being felt?)
- "What do you want to know more about? From whom? What would you want to ask?"

Eyewitness Accounts

Provide students with copies of eyewitness accounts from PBS Web site: American Experience, Surviving the Dust Bowl, People and Events. Ask photojournalists to create a found poem from one of the testimonial accounts of surviving the Dust Bowl. Read poems aloud to give voice to those "interviews."

Reading Narrative Poems from *Out of the Dust*

Ask students to read various diary entries from Out of the Dust. Introduce the poems as if they were evidence of research to give to Eleanor Roosevelt. Read poems aloud to represent the voices of the people.

- "How might you capture that scene in a photograph?"
- "Is there anything in these poems that changes your ideas about how to respond to the Presidents' requests?"

Parallel Poem: Guthrie's *Dustbowl Refugee*

Play a recording of Guthrie's *Dustbowl Refugee* (2000). Provide students with a copy of the lyrics. Ask them to write a parallel poem by first replacing the word "refugee" with a term of their own, then making whatever other changes they would like that are necessary fit the new idea related to the word they have chosen, but still maintain the rhythm and structure of the song. The poems may or may not be about the Dust Bowl. Share poems.

Optional Choral Montage: If Students Have Written a Poetic Diary Entry to Billie Jo's Journal for *Out of the Dust*

Ask students to point to a line in the poetry they have written. They are to read the line when you point to them. You orchestrate the choral response by choosing which student will speak, and how often. You may use one student's line as a refrain and repeat any response as often as you feel is appropriate.

- "Who has a line that will serve as an appropriate opening? Closing? Refrain?"

Students read lines as orchestrated.

- "What themes did you hear?" (See *Assessment Rubric for Poetry* on page 25.)

Letter of Recommendation to Mrs. Roosevelt

- "As well-informed photojournalists, what advise do you want Mrs. Roosevelt to take back to the President?"
- "How could the money best be spent?"
- "Don't tell. Write the letter of recommendation."

Read responses aloud before collecting them for Mrs. Roosevelt to take to the President. See *Assessment Rubric for Letters of Recommendation* on page 26 based on criteria set by Mehlich and Smith-Worthington (1997). See also *Assessment Rubric for Role-Play: Photojournalists in the Dust Bowl* on page 27.

Letter from Eleanor Roosevelt to the Photojournalists

The White House
Washington, D.C.
August 1936

Dear Photo-Journalists:

The President and I have admired your photographs and your writings which have been published in LIFE, THE NEW YORK TIMES, and other national and local newspapers and magazines.

You have truthfully portrayed not only the suffering but also the strength and dignity of our fellow citizens which we want you to know has inspired us to continue our work during this time of great economic depression. Thank you.

We hope you will be able to perform a service of crucial national importance. We invite you to join a PRESIDENTIAL TASK FORCE in order to document the human and environmental consequences of what people have been calling 'The Dust Bowl.'

The President is drafting a bill that will be sent to Congress before the end of the year. Your task would be to advise the President:

- Document the human and environmental consequences in photographs and writing;
- Recommend how money could best be spent to deal with these problems.

We are relying on your eyes, your ears, and your hearts to direct the President and the Congress toward action.

We want you to come to Washington, D.C., before the end of August to report directly to the President and the Cabinet.

If you feel you are able to join this project, please meet at the WPA Hall in Los Angeles, California, at 11:00 a.m. on August 15, 1936. Your reasonable expenses will of course be covered.

DO NOT TELL ANYONE ABOUT THIS MEETING. We want you to report the truth as you see, hear, and feel it.

Sincerely,

Eleanor Roosevelt

Eleanor Roosevelt

Assessment Rubric for Poetry

	Publishable (4 points)	*Excellent (3 points)*	*Revised (2 points)*	*Draft Stage (1 point)*	*Score*
Message/ Meaning (mood, key point, image)	Thought-provoking; evoked an emotional response; passionate; evocative, empathetic; experiential; intense. The message is fully developed.	Clear message focused on a central idea, feeling or experience and engages the reader. There are minor lapses in development and/or clarity.	The poem lacks focus and does not completely engage the reader. The message needs more development and clarity.	The poem does not focus on a central thought and does not engage the reader. The message is confusing.	__ x 8 = __ (32 points)
Form/ Organization	The form conveys the meaning; form provides appropriate, even unique, structure for the main idea. Poem makes effective use of poetic elements such as sound devices, rhyme, refrain, onomatopoeia, and meter.	The form is adequate in providing structure for the main idea. Poem makes use of limited poetic elements such as sound devices, rhyme, refrain, and meter.	The form is not adequate in providing structure for the main idea. It uses prose or sentence format. Little evidence of poetic elements.	Form is lacking poetic format. There is no use of any standard poetic elements.	__ x 5 = __ (20 points)
Imagery, Use of Figurative Language, Symbolism	Radiant images, sensuous, vivid use of language and poetic devices. Unexpected comparisons are made.	Clear attempt to create effective images using poetic language and devices. Some use of metaphor and/or simile.	Limited command of poetic language and devices. There are errors in the use of language.	Reads like prose instead of poetry. Denotative rather than connotative language; does not show command of basic literary devices. There are major errors in the use of language.	__ x 7 = __ (28 points)
Illustration	The illustration captures the reader's attention and enhances the meaning of the poem.	The illustration captures the reader's attention, but does not enhance the meaning of the poem.	Artistic modes are not well-developed and detract from the meaning of the poem.	The illustrations lack technical competence and detract from the message.	__ x 5 = (20 points)
			Total Points (100 possible)		
			Grade Scale	A = B = C = D =	Grade:

Assessment Rubric for Letters of Recommendation

	Exemplary (4)	*Proficient (3)*	*Developing (2)*	*Beginning (1)*	*Score*
Introduction	Clearly stated purpose; briefly explains the problem and gives a preview of the recommendation	Clear statement of purpose and rationale for the letter; gives a hint as to the recommendation	Purpose is unclear or ambiguous; no hint of the recommendation	No clear beginning	__ x 2 __ (8 points)
Recommendation	Convincing solution to the problem or need; presents persuasive arguments based on researched facts, opinions of authorities and logical thinking	Viable solution to the problem or need; persuasive arguments are somewhat convincing and based more on opinions than facts	Unreasonable, unpractical, or inappropriate solution to the problem or need; arguments lack critical thinking and logical development	No attempt to make a reasonable recommendation	__ x 8 __ (32 points)
Budget	Clearly articulated allocation of funds requested; creative and detailed request for government funds	Basic outline of allocation for funds requested; budget represents a reasonable request for government funds	Budget is either missing or is an outrageous request for government funds	No budge indicated	__ x 3 __ (12 points)
Closing	Poignant, succinct restatement of the recommendation to the President	Good nutshell statement of the recommendation to the President	Unclear, ambiguous statement of the recommendation to the President	No closing	__ x 1 = __ (4 points)
Format	Correct business letter format: heading, inside address, salutation, body, closing, signature; error-free	Mostly correct business letter format, but error-free	Incorrect business letter format and contains errors in punctuation, spelling, and/or grammar	No attempt to write a letter using correct conventions	__ x 2 = __ (8 points)
			Total Points (64 possible)		
			Grade Scale	A = B = C = D =	Grade:

Assessment Rubric for Role-Play: Photojournalists in the Dust Bowl

Role-Play Participation	*Exceptional (4)*	*Admirable (3)*	*Acceptable (2)*	*Not Yet (1)*	*Score*
Comments/ Content	Comments are frequent; informed, insightful; contributions are well-supported with detail	Comments are thoughtful logical, but not persuasive	Comments are not clearly stated and lack supporting details	Participation is lacking; no opinions are evident	__ x 5 = __ (20 points)
Connections to Classmates	Clarified information; asked questions; acknowledged others; disagreed when appropriate; entertained and developed new ideas	Demonstrated some active listening skills; encouraged others to participate; offered new ideas	Participated in the role play in a respectful manner	Socialized with the group, but didn't join in the role-play	__ x 3 = __ (12 points)
Speaking Skills	Excellent control of volume, articulation, eye contact, enthusiasm, confidence	Good control and articulation of the message. Good eye contact, but needs more confidence	Good delivery of the message. Not much confidence or enthusiasm	Not articulate, and barely audible	__ x 3 = __ (12 points)
Listening Skills	Respectful; hitch-hiked on others' ideas, used active listening skills-summarized, clarified, paraphrased	Respectful; built the conversation based on others' responses	Participated appropriately in response to others.	Interrupted others in order to speak	__ x 3 = __ (12 points)
Role-Play Skills	Stayed in character; showed imagination; creative interpretations	Stayed in character and made significant contributions	Played along for the most part	Out of character more than in character	__ x 3 = __ (12 points)
Tableaux	Gestures and body posture convey theme; evidence of planning and rehearsal; creative interpretation	Gestures and body postures give an obvious interpretation of the intended message	Gestures and body gestures portray a scene rather than an idea	Theme or event is not understood by classmates; more planning is needed	__ x 4 = __ (16 points)
Self-Reflection	Gained new knowledge and insight	Reflected on the group's participation and understanding of the Dust Bowl activities	Reflected on one's own understanding and participation	Described the events of the role-play	__ x 4 = __ (16 points)

Total Points (100 possible)

Grade Scale
A =
B =
C=
D=

Bibliography

Edmiston, B. (1998). Drama as inquiry: Teachers and students as co-researchers. In J. Wilhelm & B. Edmiston, *Imagining to learn; inquiry, ethics, and integration through drama.* Portsmouth, NH: Heinemann.

Freedman, R. (1993). *Eleanor Roosevelt: A life of discovery.* New York: Clarion Books.

Guthrie, W. (2000). *Dust Bowl refugee* (CD). New York: BMG/Buddah Records.

Hesse, K. (1997). *Out of the dust: A novel.* New York: Scholastic Press.

Manna, A. (1999). *Young adult literature course materials.* Kent, OH: Kent State University.

Mehlich, S., & Smith-Worthington, D. (1997). *Technical writing for success: A school-to-work approach.* Cincinnati, OH: South-Western Educational Publishing.

O'Neill, C. (1995). *Drama worlds: A framework for process drama.* Portsmouth, NH: Heinemann.

Porter, T. (1997). *Treasures in the dust.* New York: Harper/Trophy.

Prescott, J. (Ed.). (1995). *America at the cross roads: Great photographs from the thirties.* New York: Bromption Books.

Raven, M. T. (1997). *Angels in the dust.* Mahwah, NJ: Troll Communications.

Rylant, C. (1994). *Something permanent.* New York: Harcourt Brace.

Short, K., Harst, J., & Burke, C. (1996). *Creating classrooms for authors and inquirers.* Portsmouth, NH: Heinemann.

Stanley, J. (1992). *Children of the Dust Bowl.* New York: Crown.

4

READERS' THEATER FOR *DELIVER US FROM EVIE*

Jacqueline Glasgow

Everyone needs to talk-to hear and to play with language, to exercise the mind and emotions and tongue together. Out of this spirited speech can come meaningful, flavourful language, worth the time and effort of writing and rewriting, phrasing, rehearsing, and reading aloud.

Wolsch and Wolsch, 1988

Overview of Authentic Learning

Readers' theater provides a way to involve students in presenting literature to other students in dramatic form. McCaslin (1990) defines readers' theater as the oral presentation of drama, prose, or poetry by two or more readers. Readers' theater differs from orally reading a selection in that several readers take the parts of the characters in the story. Instead of memorizing or improvising their parts as in other types of theater productions, the players read their lines. Since the emphasis is on what the audience hears rather than sees, selection of the passages and dramatic reading of the script is very important. Members of the audience must use their imagination to visualize what is going on, because movement and action are limited. Although there is no one correct arrangement of the cast in presenting a readers' theater, an effective procedure is to have the student readers stand in a line facing away from the audience and then turn toward the audience when they read their part. As with oral reading of literature, it is important before the presentation to convey to the audience an introduction and the setting, conflict, and mood. Readers' theater scripts generally contain a great deal of dialogue and often are adapted from a piece of literature by the teacher or the students themselves. According to Lois Walker (2001), readers' theater is a reading resource, a teaching tool, a performance vehicle, and a chance to play with language. Readers' theater assists students in gaining a conceptual understanding of the theme that would not be achieved by them if they read without the support that drama affords.

According to Shepard (1996), there are many styles of readers' theater, but nearly all share these traits:

- No full memorization. Scripts are held during performance.
- No full costume. If used at all, costumes are partial and suggestive, or neutral and uniform.
- No full stage sets. If used at all, sets are simple and suggestive.
- Narration provides the framework for dramatic action.

The following assessment is one way to prepare students to write and perform their own scripts.

NCTE/IRA Standards

In preparing students to write and perform readers' theater, students will:

- Write a letter discussing a conflict with their mothers. In doing so, students will adjust their use of spoken, written, and visual language to communicate effectively with a variety of audiences and for different purposes (NCTE/IRA Standard #4), and they will employ a wide range of strategies as they write and use different writing process elements appropriately to communicate with different audiences for a variety of purposes (NCTE/IRA Standard #5).
- Create and present tableaux (freeze frames). In doing so, students will participate as knowledgeable, reflective, creative, and critical members of a variety of literacy communities (NCTE/IRA Standard #11), and students will use spoken, written, and visual language to accomplish their own purposes (NCTE/IRA Standard #12).
- Select a book, write a poem in two voices, and create a readers' theater script. In doing so, students will adjust their use of spoken, written, and visual language to communicate effectively with a variety of audiences and for different purposes (NCTE/IRA Standard #4); they will employ a wide range of strategies as they write and use different writing process elements appropriately to communicate with different audiences for a variety of purposes (NCTE/IRA Standard #5); and, they will apply knowledge of language structure, language conventions, figurative language, and genre to create, critique, and discuss texts.
- Perform a prepared script. In doing so, students will participate as knowledgeable, reflective, creative, and critical members of a vari-

ety of literacy communities (NCTE/IRA Standard #11), and students will use spoken, written, and visual language to accomplish their own purposes (NCTE/IRA Standard #12).

Performance Tasks for Readers' Theater

A Letter of Complaint

Ask students to brainstorm a list of conflicts or disagreements they have had with their mothers. Then ask students to take on the role of victim in a particular incidence and write a letter about this situation to someone who they feel could advise them or help them. This brief frontloading activity gets students thinking about injustice and conflict in personally relevant terms and accesses background knowledge and opinions that can help them relate to the action and themes of the story. It also works to challenge their understanding of injustice and consider ways to resolve conflict.

Tableaux Drama/Slide Show

In tableaux drama, students create a frozen picture with their bodies. The picture can be of an event, a scene, or a conflict from the story. Pictures of missing or implied scenes can be created to help them fill story gaps. A variation of the tableau is to ask students to perform a brief role-play and then freeze it at the climax. At other times, students like to freeze, role-play, freeze again, and so on in a kind of slide show. Another variation is to tap the shoulder of a student in a freeze and have them "come to life" to report on their feelings at the time of the picture. In this case, divide the students into groups of four or five and ask them to create a tableau for a scene where Evie is in conflict with her mother. Ask each group to perform the freeze with other students walking around the group and suggesting interpretations of what they see.

Poem in Two Voices

Ask students, now back in their groups, to find a passage in the text that best shows the conflict between Evie and her mother. The passage may be the same one that they just performed. Ask them to write a poem in two voices taken from that scene in the style of Paul Fleischman's (1988) *Joyful Noise: Poems in Two Voices*. There is more than one way to compose a poem for two voices as a way to respond to fiction; for instance, two characters from the same book or story can speak to each other about a critical event or situation they have in common, giving two points of view. Sometimes, one voice will speak alone; at other times, they both speak together. Two people or two groups of people are needed to read the poem; each person or group reads one column. See *Poem in Two Voices "After the Storm"* on page 32 written about the mother/daughter conflict in Kerr's (1995) *Deliver Us From Evie.*

Poem in Two Voices "After the Storm"

Evie	*Mom*
I love to read	I love to read
Books, lives, hearts	Books, lives, weather.
Do you like my new scarf?	It didn't come from around here.
Do you like my bracelet?	What boy gave that to you?
Do you want to read what's written on the back?	Not now, the weather's turning bad and I've got mending to finish.
How do you like my new heart's happiness?	It didn't come from around here.
Do you want to read what's written on the front?	Not now, but I will after the storm's over.

See *Oral Presentation Assessment Rubric* on the following page that can be used to assess the poem in two voices and/or the readers' theater.

Selecting a Book

Ask students to work collaboratively in selecting a novel for a readers' theater based on one of the books they have read by M. E. Kerr or other authors. Use the following checklist from Baie Comeau High School to help select the book:

- two or more characters interact
- lots of dialogue or indirect speech to turn into direct speech
- many events occur in a short space of time
- the personality of the characters(s) is developed in the story
- the story makes sense and will lend itself naturally to creating a script

Preparing the Script

For *Deliver Us From Evie,* one approach to preparing a script would be to divide the class into two sections: one to represent Evie, and one to represent her mother, Cynthia. Each section should be divided into various chapters of the book to include many of the conversations in the script. Ask each student in the Evie group to find lines from the text that Evie said to her mother and write them on an index card, noting the page number. Then ask the Cynthia group to do the same thing about the lines that Evie's mother said to Evie. Every student should have a line to present to the class, and some may have the same line as someone else. Ask one student to write a line for Parr introducing the conflict, and ask another student to write a line for Parr concluding the script. When

Oral Presentation Assessment Rubric

	Exemplary (4)	*Proficient (3)*	*Developing (2)*	*Beginning (1)*	*Score*
Vocal Projection	The speaker delivers the message in a lively, enthusiastic fashion; uses voice to add emphasis and interest	The speaker enunciates clearly, uses expression, and is clearly heard by the audience	The speaker speaks clearly, but with little expression	The speaker does not enunciate and is difficult to hear	__ x5 = __ (20 points)
Pacing	The speaker reads with good use of pause, giving sentence drama	The speaker's pattern of delivery is generally successful but may have uneven patterns	The speaker's delivery is either too fast or too slow	The rate is so fast that you cannot understand most of the message	__ x 5 = __ (20 points)
Oral Reading Miscues (Errors)	The speaker reads fluently without making errors	The speaker reads fluently without making errors that change the meaning of the text	The speaker reads with miscues that change the meaning of the text	The speaker is unable to pronounce many of the words	__ x 5 = __ (20 points)

Total Points (60 possible)

Grade Scale
A =
B =
C =
D =

everyone is ready, begin a "jump-in" reading of the lines. Begin with Parr's introduction and ask students to "jump in" with their line when they feel it is appropriate. Alternate groups so that first someone reads a line for Mother and then someone else reads a line for Evie. Some lines may be repeated, which will have a chorus or refrain effect. When everyone has performed his/her line, the person with Parr's conclusion should finish.

Writing the Script

Another approach to this assignment is to ask students to experiment turning the first few pages of the book into a script. Practice turning indirect speech (Evie was feeling really tired and hungry) into direct speech ("I feel really tired and hungry"). Then, continue writing the script (see *Assessment Rubric for Readers' Theater Script*). Use the following guide prepared by Baie Comeau High School as an aid:

- Character's name goes on the left-hand side in bold print.
- Short dialogue is used to keep the story moving quickly.
- Narrator ties the story together or explains reasons for events. Do not overuse this character.
- Narrator 1 and Narrator 2 may be better if there's lots of description, details.
- Sound effects, simple props or music enhance a Readers' Theater.
- A chorus of several voices gives dramatic effect.
- The use of rhyme, repetitive structure, internal rhyme, dramatic excitement and heaps of action also enhance a script.

Reading a Prepared Script for Readers' Theater

After duplicating classroom scripts and passing them out to all students, assign parts to various class members. Ask them to highlight the parts they will read and give them a chance to read the script silently before performing it. Ask readers to assemble chairs in the front of the room for the performance. When the reading is completed, discuss the story, the reading, reassess parts, swap scripts, and read again. See the following readers' theater script based on the mother/daughter conflict in Kerr's *Deliver Us From Evie* in *Sample Readers' Theater from Deliver Us From Evie* on page 36.

Readers' theater can be further enhanced by playing music softly in the background and by dimming the lights so that the students' full attention is in listening and experiencing the emotional impact of the theater experience. After a few moments of silence, divide students into discussion groups to debrief and talk about the ideas that emerged from the readers' theater. In this case, the Evie group should discuss Evie's struggle and courage in "coming out" to her mother and family with her sexual identity. What were the consequences she faced? The Cynthia group should discuss Cynthia's attitudes in coming to grips

Assessment Rubric for Readers' Theater Script

	Distinguished (4)	*Proficient* (3)	*Apprentice* (2)	*Novice* (1)	*Score*
Dialogue	Creative use of dialogue to advance the drama, add humor, make story more realistic, create mood	Appropriate use of dialogue to keep the drama moving	Dialogue is slow-moving, limited, and stereotypical	Dialogue doesn't make sense or develop the drama	__ x6 = __ (24 points)
Characters	Balance of characters and voices; multiple perspectives portrayed	Appropriate balance of characters and voices; differing perspectives portrayed	Imbalance of characters and voices; limited perspectives portrayed	Some characters do all the talking; only one point of view portrayed	__ x6 = __ (24 points)
Narrative	Much more dialogue than narrative; creative use of narrator	Balance of dialogue and narrative; good use of narrator	More narrative than dialogue; narrator explains too much	Much more narrative than dialogue; narrator has the major role	__ x4 = __ (16 points)
Language	Rich use of figurative language, rhyme, repetitive structure, internal rhyme, dramatic excitement	Good use of dramatic techniques	Limited use of dramatic techniques	No use of dramatic techniques	__ x3 = __ (12 points)
Impact	The drama makes a significant impact on the audience; keeps listener's interest peaked throughout	The drama makes an important point; keeps listener's interest	The drama makes an ambiguous or stereotypical point; listener has difficulty maintaining interest	The drama is ambiguous and confusing; listener loses interest	__ x3 = __ (12 points)
Format and Correctness	Title page, character designations, pages numbered, clear stage directions, single-spaced block with double between speeches, mechanical correctness	Mostly correct spacing and format; mechanical errors do not distract from the reading of the script	Errors in spacing, character designations, and format; errors detract from the reading of the script	Character designations are confusing; no stage directions; incorrect format; many mechanical errors	__ x2 = __ (8 points)
Special Effects	Uses sound effects, simple props or music	Uses some special effects	Uses limited special effects	Uses no special effects	__ x1 = __ (4 points)

Sample Readers' Theater from *Deliver Us From Evie*

Parr:	"Sometimes I'd look at my mother and wonder how she'd ever brought someone like Evie into the world...the only thing they had in common was a love of reading." (p. 3)
Cynthia:	"It'd fit you, Evie!" (p. 5)
Evie:	"It might fit me but it doesn't suit me!" (p. 5)
Cynthia:	"Evie, it's not appropriate to wear to the Duffs'" (p. 6)
Evie:	"Only difference between a pigeon and a farmer today is a pigeon can still make a deposit on a John Deere tractor." (p. 7)
Cynthia:	"Where's your shirt, Evie" (p. 11)
Evie:	"Patsy lent me this... her school shirt" (p. 11)
Cynthia:	"I thought she went to Appleman Academy." (p. 11)
Evie:	"She does. But the students call it Appleperson Academy, for fun ... You can't be sexist." (p. 11)
Cynthia:	"If you'd go to Missouri, you'd automatically be a Pi Phi, Evie." (p. 23)
Evie:	"Yeah, they'd be tickled to see me clumping up their front sidewalk, wouldn't they?" (p. 23)
Cynthia:	"Evie, honey, you could be every bit as pretty as any one of the Pi Phis... if you'd just let me help you with your clothes, if you'd just change you hair, style it!" (p. 24)
Evie:	"I'm the way I am." (p. 24)
Cynthia:	"Don't comb it back, Evie!" (p. 28)
Evie:	"I'm the way I am." (p. 24)
Cynthia:	"Whew! Evie! The smoke from your cigarettes is too heavy in here." (p. 33)
Evie:	"I'm the way I am." (p. 24)
Cynthia:	"Khaki trousers to a concert? Evie, you have to wear a skirt!" (p. 36)
Evie:	"I'm the way I am." (p. 24)
Cynthia:	"You're what everybody's always thought one of those women was like" (p. 67).
Evie:	"I'm the way I am." (p. 24)
Cynthia:	"Who's Jane Doe, honey?" (p. 83)
Evie:	"It's me. Okay?...am I forced to go around sneaking to spare your feelings?" (p. 84)
Cynthia:	"Evie, I wasn't born yesterday. I'm not unfamiliar with lesbianism. Gays. Whatever you call it. Is that what you claim you are?" (p. 85)
Evie:	"I'm the way I am." (p. 24)
Cynthia:	"You don't know that for sure, honey." (p. 85)
Evie:	"Some of us look it, Mom! I know you so-called normal people would like it better if we looked as much like all of you as possible, but others go for the ones who don't, can't, and never will!" (p. 86)
Cynthia:	"Don't tell your father about this, Evie." (p. 87)
Evie:	"Lie?" (p. 87)
Cynthia:	"Yes. Lie." (p. 87)
Evie:	"Far as I know, there's no law against two females seeing each other." (p. 128)
Cynthia:	"How'm I supposed to take it?" (p. 129)
Evie:	"I am what I am, and whatever that is, it hasn't bothered anyone before, has it?" (p. 130)
Cynthia:	"Be careful of the friends you make, too, Evie. Don't seek out the gays. Think about that lifestyle, Evie...Why don't you see a doctor, honey? There are clinics in a big city like New York." (p. 145)
Evie:	"I'm the way I am." (p. 24)
Cynthia:	"Thanks for the nightie, honey. I never had anything from Paris, France...did you buy yourself some new clothes?" (p. 176)
Evie:	"I didn't buy myself a nightie." (p. 176)
Cynthia:	"So long, you two. Don't you two be strangers!" (p. 176)
Parr:	"Evie can't pass herself off as something else. It isn't in her nature." (p. 65)

with her daughter's relationship. How was she supposed to deal with the shock, despair, and damage to the family? After the class discussion, students could locate either Internet or local support groups that offer hope and healing for parents, families, or children in need of help or restoration. Students can write editorials or newsletters to show how their attitudes and perspectives changed during the course of this study.

Bibliography

Baie Comeau High School (October 31, 2001). *Reader's theatre* [Online]. Available: http://www.qesn. meq.gouv.qc.ca/schools/bchs/rtheatre/teach2.htm.

Fleischman, P. (1988). *Joyful noise: Poems in two voices.* New York: Harper Trophy.

Kerr, M.E. (1996). *Deliver us from Evie.* New York: Harper Trophy.

McCaslin, N. (1990). *Creative drama in the classroom* (5th ed.). White Plains, NY: Longman.

Shepard, A. (1996). *Aaron Shepard's RT page* [Online]. Available: http://www.aaronshep.com/rt/wahtis.html.

Walker, L. (2001). *Readers' theatre reading resources and teaching tools children will love* [Online]. Available: http://loiswalker.com/catalog/teach.html.

Wolsch, R. A., & Wolsch, L. A. C. (1988). *From speaking to writing to reading: Relating the arts of communication.* New York: Columbia University Teachers College.

Part II

Multimedia Assessments

5

CULTURES AND GENERATIONS: RESEARCHING THE FAMILY HISTORY

Linda J. Rice

You are a very special person, individually conceived and created.
However, you also belong to a history, to a people, and to places, and have
been influenced by ideas, views, hopes, longings, failures, and successes.
People the world over who are mentally healthy know a lot about their past.
They build on that past using it for succor as Wordsworth did his memory
of Tintern Abbey.
As these healthy people peer into the past,
they learn to respect all peoples and their histories and cultures.
This document, your research into your identity, will be a lasting record
that will inform even...
offsprings.

Tchaiko Kwayana

Adapted from Tchaiko Ruramai Kwayana's article in January 1996 English Journal, pp. 62–66.

Overview of Authentic Learning

Kwayana's words sum up the basic rationale for this year-long I-Search, research, and writing project: to look to the past and see how our unique history, culture, and family experience has shaped who we are. A further aim is to involve parents and guardians in our educational endeavors. This unit is all about making connections, through cooperation with the family and substantial reading of ethnic tales and geographic histories.

The culminating experience of the four-part paper is its presentation, where parents/guardians (relatives and friends) join their children during class or on scheduled evenings to share part of their family history.

Throughout this project, students develop their skills in interviewing, researching, organizing information, integrating sources, writing, and presenting. In its entirety, this project is a microcosm of the five language arts strands (reading, writing, speaking, listening, viewing).

NCTE/IRA Standards

In this unit students will:

- As they explore their family heritage, read a wide range of print and nonprint texts to build an understanding of texts, of themselves, and of cultures outside the United States (NCTE/IRA Standard #1). In addition to helping students learn vital and important information, this exploration will also provide them with personal fulfillment as they learn more of their own history. Fiction, nonfiction, film, music, and art combine to meet this standard.
- Conduct research, using a variety of technological and information resources, read a range of literature that spans genres and time periods, and synthesize these findings to identify themes and issues prevalent in the culture of their ethnicity and tradition (NCTE/IRA Standards #2, 7, & 8). This also assists students' understanding of their connectedness to broader "families."
- Generate questions, draw on their prior experience, and interact with family members as a way of appreciating the "texts" (NCTE/IRA Standards #3 & 7) of their family's oral language stories. Students will then transform these oral narratives into parts of their own papers and presentations for the unit.
- Adjust their use of spoken language (in interviews) and written language (in reporting their findings) to communicate effectively with different audiences (NCTE/IRA Standard #4) including family members, visitors (on presentation days), peers, and the teacher. In light of this, however, it should be noted that students are to prize and preserve the authentic voice of the story-teller, wherever possible, thus developing an understanding of, and respect for, diversity

in language use, patterns, and dialects across cultures, ethnic groups, and geographic regions (NCTE/IRA Standard #9).

- Employ, through the four components of the paper that culminate in a presentation, a wide range of strategies for writing, revising, and communicating for a variety of purposes (NCTE/IRA Standard #5) such as storytelling, reporting research findings, and projecting into the future (through an epilogue). Embedded in this process is students' application of knowledge of language structure, language conventions, and genre that facilitate their communication of themes from both print and nonprint texts (NCTE/IRA Standard #6).
- Use spoken, written, and visual language (NCTE/IRA Standard #12), particularly in the presentation component where parents/guardians join the class to share some aspect of their family's unique culture and/or heritage.

Two Approaches for Teaching this Unit

Short-Term, Intensive Focus

One approach is to have this unit as the central one for a nine-week grading period. In that case, this would be all students would work on (which is plenty because the unit is comprehensive in terms of language arts skill-building: reading, writing, listening, speaking, and viewing).

Long-Term, Extended Focus

The other approach (which is the way the unit is presented here) is to carry the project throughout the year, with one of the four parts due each nine weeks. This approach allows more time for students to conduct interviews and really start to enjoy learning from their relatives. It also facilitates students' efforts to find, and when necessary, share resources, including correspondence from genealogical societies students may write to for information. Finally, if taken as a year-long endeavor, students leave the project with the satisfaction of having created what is, in effect, a four-chapter book, a mark of pride for themselves and their family.

Major Assessment Components

Performance Task for Family Interviews

Part I (first nine weeks) is based on family interviews (teachers should steer students away from tracing the family's origin back to another continent for Part I because that will be the entire focus of Part III of the paper) and should be recorded in the form of a series of narratives that address the following:

- Autobiographical incidents (funny stories, how grandma met grandpa, hobbies, etc.)
- Special family celebrations
- Ethnic maxims and proverbs (e.g., why grandma always says "trust in God, but tie your camel")
- Foods or other distinct family experience and behavior
- Sketches of various family members, heroes, heroines, folk personalities, and "characters"

At the beginning of Part I, teachers should prepare one or two mini-lessons and practice sessions on interviewing (see *Helpful Hints for Successful Interviewing*). Then, after having two to three weeks to conduct several in-depth interviews with family members, students will seek to synthesize and/or retell stories representative of their lives and heritage in the form of an I-Search paper. Of the four parts of the total paper, Part I is typically the longest because students gather so many interesting stories through the interviewing process. Five pages would be a minimum, but it is not uncommon for highly motivated students to write papers as long as 20 to 30 pages. One of the particular beauties of the interview process is that much of the paper seems to "write itself" as students serve as editors and lenses through which the stories are synthesized and retold. Students should include a Works Cited page that includes who they interviewed and when. Using MLA for all style specifications (header, spacing, pagination, parenthetical documentation, Works Cited, etc.) works well.

Helpful Hints for Successful Interviewing

- Arrive prepared! Write 10 to 25 questions before conducting the interview.
- Ask open-ended questions that begin with "Tell me about..." or "Describe..."
- Attempt to go back as many generations as possible. Keep in mind that if you interview your grandparents and ask what they remember about their grandparents (or better yet what they remember their grandparents saying about their grandparents) you will be able to go back four (or more) generations quite readily.
- Write only one or two questions per page, leaving plenty of room for taking notes.
- Audio taping interviews is fine, but does not replace note taking. Any good researcher knows the foibles of batteries running out, technical problems, etc.

- Take detailed notes (to be collected for a grade) as evidence of interviewing/listening skill.
- "Preserve the language" of the speaker. Even if the language isn't entirely "proper" or "standard English," it should be preserved (where possible) to convey the speaker's sense of voice and ownership of words and language. It is often the interviewee's dialect or way of "turning the phrase" that gives Part I its rich texture, reflecting the uniqueness of the student's own family.

Establishing checkpoints (see *Checking to Ensure Quality Interviews* on page 45) to ensure that the interviews take place in a timely fashion is virtually essential for high-quality outcomes. Such benchmarks in timing promote careful planning, rather than rushing to meet final deadlines. And the longer the students have to process the interviews and incorporate them into Part I of the total paper, the more likely they are to truly find connectors, synthesize, and produce a high-quality final draft that they want to share with relatives. See *Assessment Rubric for Part I: Family Interviews* on page 47.

Checkpoints to Ensure Quality Interviews

Week 1

- Mini-lesson: "Helpful Hints for Successful Interviewing"
- Write questions and practice interviewing (optional)
- List of people to be interviewed (plus a list of questions that will be asked, realizing that these will vary depending on whom is being interviewed)

Week 2

- Completion of 1st interview (as evidenced by turning in notes, tape, or transcript)

Week 3

- Completion of 2nd interview (with supporting evidence)

Week 4

- Completion of 3rd and 4th interviews (with supporting evidence)

Week 5

Progress Report/Early Draft of Part I

- Write a Progress Report (3 pages double-spaced and typed, MLA style) that demonstrates ability to transform and integrate information from interviews into students' own paper, Part I

- The Progress Report should include the following:
  - Results of at least two interviews
  - Parenthetical documentation to cite the information in the paper
- On the day the Progress Report is due, each student should select a portion of his/her paper to read aloud to the class so students share/gain ideas on how to put the paper together
- Use the Progress Report day for question-answer time

Week 6

Rough Draft

- Same procedure as Progress Report/Early Draft of Part I, but this time at least 5 pages and transforming/integrating information from all four interviews

Week 7 or 8

Final Draft Due

- Naturally, collecting earlier drafts to see what kinds of revisions students made is helpful to gain insight into various elements of the process that may need additional emphasis in the future

Performance Task for Surveying the Larger Cultural/Ethnic Family

Whereas Part I focused on the stories, gathered through interviews, of the students' immediate families, Part II extends to the students' broader cultural/ethnic family. This search may begin in the school's library/media center but will likely continue at a local, regional, college, or university library, online, or through other reference sources available to the students. Here, students examine the literature, history, art, and music of their culture/ethnicity. The challenge here is to originate the focus of all such study in a country other than the United States—unless the student is Native American, although here, too, the study may begin away from this country.

After students have conducted their research survey, they will begin the composition for Part II. This will resemble a "traditional" research paper in that it will report the findings of the research survey. Guiding questions to assist students in writing the paper for Part II are as follows:

- What themes run through these works/reflections of culture/ethnicity?
- What concerns and interests seem to be the focus of the various writings, films, songs, and works of art?
- What thematic and other shifts in the writings of the ethnic group have occurred over time? How has the group changed?

Assessment Rubric for Part I: Family Interviews

	Exemplary	*Acceptable*	*Needs Improvement*
Preliminary List	Includes names and relationships of 4 family members to be interviewed followed by questions for each; questions are predominantly open-ended	Includes only 3 people, but otherwise matches standard for exemplary performance; or missing relationships; or having only 2 to 3 open-ended questions	Includes only 1 or 2 people; missing relationships and/or only has 1 to 2 open-ended questions
Interview 1	Comprehensive notes and/or transcript neatly prepared	Sketchy notes or difficult to read due to poor organization or neatness	Only a tape
Interview 2	Comprehensive notes and/or transcript neatly prepared	Sketchy notes or difficult to read due to poor organization or neatness	Only a tape
Progress Report/Early Draft	At least 3 pages; Integrates results of at least 2 interviews; Includes parenthetical documentation in MLA style; shared with the class	At least 2 ½ pages; Includes results of two interviews, but not in a way that indicates that they were "processed" by the writer (perhaps needing more of an introductory context or transitioning); or errors in parenthetical documentation; or chose not to share with the class	2 or fewer pages or includes findings from only one interview
Interview 3	Comprehensive notes and/or transcript neatly prepared	Sketchy notes or difficult to read due to poor organization or neatness	Only a tape
Interview 4	Comprehensive notes and/or transcript neatly prepared	Sketchy notes or difficult to read as a result of poor organization or neatness	Only a tape
Rough Draft	At least 5 pages, prepared MLA style; Integrates results of at least 4 interviews; Includes parenthetical documentation in MLA style	At least 4 pages; includes results of only 3 interviews, or all 4, but not in a way that indicates that they were "processed" by the writer (perhaps needing more of an introductory context or transitioning); or errors in parenthetical documentation	3 or fewer pages or includes findings from only one or two interviews
Final Draft	Same criteria as Rough Draft, only now prepared with MLA header and Works Cited page; 2 or fewer errors per page in grammar, usage, and/or mechanics	Same criteria as Rough Draft, only now prepared with MLA header and Works Cited page; Contains 3 to 4 errors per page in grammar, usage, and/or mechanics	Same criteria as Rough Draft, problems with MLA header and/or Works Cited page; contains 5 or more errors per page in grammar, usage, and/or mechanics

In exploring these questions, students should include descriptions and samples of the literature, including several of the following genres: folktales, myths, legends, proverbs, essays, historical fiction, nonfiction, short stories, novels, poetry, and drama. And, to gain a broader perspective on culture, students will also reference a film, a song, and a work of art representing the culture and/or ethnicity of their family. In their writing, students are encouraged to work on transitional sentences and integration of various resources. As a guideline, students might be asked to include three different literary genres and references to a film, an opera, and a painting—seeking out the similarities, juxtapositions, and/or defining aspects of culture. Five to ten pages is an average requirement for Part II's final paper, which should also demonstrate proper MLA style, parenthetical documentation, and Works Cited. See *Assessment Rubric for Part II: Survey of the Larger Cultural/Ethnic Family* on page 49.

Performance Task for Odyssey of the Family

Part III (third nine weeks) traces the odyssey of the extended family from some other country or continent to the United States and finally to the student's own city or town. Here, students sharpen their awareness of geography and seek demographic information reflecting the place(s) from which their ethnic group came. The work of historians, anthropologists, and archaeologists may be helpful in mapping the demographic changes. The reasons families immigrated are of particular interest (often with historical connections). For example, the rise of Hitler drove some out of Germany, and the potato famine prompted the movement of the Irish.

Part III combines traditional library-based research and I-Search strategies as students again return to their relatives to seek information about the family's odyssey. In the event that they are not able to trace the family line back to another country or gather information on the voyages of former generations, students may focus this section on descriptions of places (homes or even regions) where the family has lived in more recent years. Maps (and photographs of family homes) are often included in the appendices of Part III, providing evidence of students' effort to trace the family's journey and learn about geography in the process. Because the presentations (collaboratively prepared by students and parents) occur during the third nine weeks, this portion of the paper tends to be relatively short—four to seven pages. See *Assessment Rubric for Part III: The Odyssey of the Family* on page 50.

The Presentation Component (Third Nine Weeks)

While the ongoing behind-the-scenes role of parents/guardians is to help direct students in their quest to find information and family memorabilia, their visible role will appear in the third nine weeks when class time and evening presentations of family histories are scheduled. Ideally, for the presentations (to last 20 to 25 minutes each), parents/guardians will work collaboratively with their son/daughter. This is a fun time of sharing that often includes family photos, videos, maps, artifacts, heirlooms, and scrumptious ethnic dishes! It is

Assessment Rubric for Part II: Survey of the Larger Cultural/Ethnic Family

	Exemplary	*Acceptable*	*Needs Improvement*
Required Genres	3 genres of literature 1 film 1 song 1 work of art	Missing any one of those listed in the exemplary column	Missing 2 or more
Integration of Sources	Woven artfully to make cross-comparisons and find similarities, themes, and logical links among the various works	Some connections seem forced or illogical; quotes/excerpts included in the paper do not work cohesively to support the intended theme	Works are written independently of one another without convincing thematic link or juxtaposition
Transitional Sentences	Evident throughout the paper; effectively bind ideas either by similarity or reasonable juxtaposition	2 to 4 instances where ideas are not linked, thus detracting from the paper's optimal build, unity, and adherence to theme	5 or more instances where transitions are needed but not present
MLA Style	Complete adherence	1 to 2 errors	3 or more errors
Length/ Development	At least 5 pages that are well-developed with interesting information and examples of the required genres, followed by thematic links and commentary that demonstrate critical synthesis	4 or 5 pages, but with some information that has limited relevance or compelling interest—in other words, there is a sense that the extra page was written not to convey the best investigation of the larger cultural/ethnic family, but to meet a page requirement	3 or fewer pages
Grammar, Mechanics, and Usage	1 or no errors per page	2 to 3 errors per page	More than 3 errors per page

Assessment Rubric for Part III: The Odyssey of the Family

	Exemplary	*Acceptable*	*Needs Improvement*
Content	Establishes connections between history (global or family) and what motivated the movement from place to place; organized to present a smooth, logically flowing picture of the family's odyssey	Establishes connections between history (global or family) and what motivated the movement from place to place; organization is weak, wandering, or appearing random in places	Tells about different places the family has lived over generations, but does not explain reasons for the migration in connection with family story or historical happenings, where relevant
Research Techniques	Includes both I-Search, as evidenced through citations of family interviews, and traditional research that establishes a kind of historical reasoning or description of various time periods at the time of various family moves (or "stays")	Includes both techniques, but with a tendency to dwell on one area of history in a way that detracts from the central family story; in other words, tendency to become too much like a book report on history, rather than a narrative of a family's journey	Based solely on family interviews with no supportive investigation (from traditional research sources) to explain the history or times in which various generations of the family lived
MLA Style	Complete adherence	1 to 2 errors	3 or more errors
Length/ Development	5 to 7 pages that are well-developed with interesting information and examples of the family's odyssey mixed with historical influences and/or time period characteristics	4 or more pages, but with some information that has limited relevance or compelling interest; in other words, there is a sense that the extra page was written not to convey the best exploration of the family's odyssey, but to meet a page requirement	fewer than 4 full pages
Grammar, Mechanics, and Usage	1 or no errors per page	2 to 3 errors per page	More than 3 errors per page

important to write a letter explaining the project and emphasizing that sharing can be as "personal" or "impersonal" as the participants are comfortable with. For example, whereas some families may bring grandparents, aunts, uncles, parents, and siblings to tell lots of stories, share photos, and recipes, others may avoid family stories altogether and instead focus on the history, literature, or art of a country of their origin.

Because of the tenuous nature of working with parents/guardians, and the necessity of allaying concerns that they, themselves, are being "assessed" in any way, full credit will be given to all students who participate in this aspect of the unit. The point is to encourage and support the involvement of the family in the learning process. Naturally, it is a kind and important gesture to write personal thank yous to the families who participate. This can also be a great PR move on behalf of the district. And the news media may even be willing to come in and "cover the story."

One may logically ask why the Presentation Component, considered to be the culminating experience of this unit, is not postponed until after the Final Collected Manuscript is completed. The reason for this is threefold. First, by the third nine weeks, students have plenty of detailed and "fresh" information to present without being overwhelmed by the additional details of Dedication, Gratitude, Table of Contents, Epilogue, and Appendices. Second, Part III of the paper (the Odyssey of the Family) tends to be shorter than the remaining three portions of the paper; therefore providing students with more time to dedicate to preparing the presentation. Third, and very practically, the end of the school year tends to have more scheduling conflicts that can detract from the positive momentum of the project. Additionally, the epilogue, in some cases, can be almost depressing as students observe the decline of their cultural/ethnic practices in the United States. Therefore, presenting in the third nine weeks is optimal.

Performance Task for the Epilogue and Final Collected Manuscript

Having completed their study of the past (through family interviews, broader ethnic/cultural research, and tracing the family's odyssey), students are now ready to take a personal look into the future. Part IV (fourth nine weeks) is a critical assessment of the strengths and weaknesses of the students' cultural/ethnic group that is to be written in the form of an epilogue. Here students pen their goals, ambitions, reservations, hopes, and predictions. Guiding questions to assist students with composing the epilogue are:

- What has happened in the case of my ethnic family?
- Have they "melted?" At all? In part? Why? Why not?
- What role can I play in adding to the esteem of my people and to society as a whole?

Also, during the final nine weeks, students will prepare the following as components of the final collected manuscript:

- Cover (must include student name, course name, date, and a title that reflects the student's unique angle on the paper as a whole) and Binder (preferably three-ring binder)
- Dedication page for the person/people the student wishes to celebrate, honor, or memorialize with the collected manuscript
- Gratitude page, a statement of appreciation for those who helped with the project
- Table of Contents
- Section dividers for Parts I–IV and Appendices

See *Assessment Rubric for Part IV: The Epilogue and Final Collected Manuscript* on page 53.

Assessment Rubric for Part IV: The Epilogue and Final Collected Manuscript

	Exemplary	*Acceptable*	*Needs Improvement*
Content (of Epilogue)	Demonstrates evidence of reflection on the central questions and shows an understanding of the epilogue as a genre	Has a one-sided view that illuminates strengths or weaknesses, but not both; otherwise demonstrates understanding of epilogue and reflection based on guiding questions	Written as a conclusion to the paper, but not in a way that is personal in its look at goals, ambitions, reservations, hopes, and predictions about family culture/ethnicity
MLA Style	Complete adherence	1 to 2 errors	3 or more errors
Grammar, Mechanics, and Usage	1 or no errors per page	2 to 3 errors per page	More than 3 errors per page
Cover and Binding	Labeled in a neat and artful manner with student's name, course name, date, and a title that reflects a unique angle on the paper as a whole; fitting binder, not a mere staple	Missing one of the required pieces of information; otherwise meets criteria of exemplary	Missing 2 or more pieces of required information or not bound except by a staple or paper clip
Dedication	Full credit if included	(See "exemplary" note)	None
Gratitude	Includes brief explanation of how those listed contributed to the project	Includes a list of names, but without reasons or explanations about how each contributed	None
Table of Contents	Professional visual appeal; consistent with the actual paper's arrangement; complete; includes titles and page numbers	1 to 2 incongruencies between what is listed and how things are actually arranged in the final collected manuscript; or problem with alignment of columns	3 or more incongruencies between what is listed and how things are actually arranged in the final collected manuscript
Section Dividers	Labeled and properly installed before each section's content	Not labeled or installed improperly	None
Appendices	Organized consistently with the way they are alluded to in the four sections of the paper; labeled	1 to 2 incongruencies between order arranged here as compared to how they are alluded to in the paper; or not properly labeled	None

6

INTERACTIVE NOVEL GROUP PRESENTATIONS

Linda J. Rice

The only good is knowledge and the only evil is ignorance.

Socrates

Overview of Authentic Learning

This unit is designed to give students a "cultural literacy" understanding of four to six "classic" American novels while reading only one from beginning to end. This is achieved as students work cooperatively to prepare "teaching presentations" to inform the rest of the class (organized in small groups to study other assigned novels) of the "basics" of their assigned novel. The goals of this unit are threefold. First, students are in regular communication to become "experts" on their assigned novel. Second, students hone their presentation skills as they "teach" their novel to the rest of the class. Third, students see their classmates as educational partners from whom they can learn.

Through the process of preparing "teaching presentations," students negotiate meaning, organize information, and seek ways to make that information interesting and relevant to their peers. "Ownership" is a key component of this unit as students become active participants in determining what the "essential elements" are, rather than passive recipients of what the teacher has to say about the novels under investigation.

In their groups, students also negotiate who will present what. This opens the door for students to discuss and share their skills (organizational/written for handout, artistic for poster, verbal for oral interpretation, and final project utilizing students' multiple intelligences).

NCTE/IRA Standards

In this unit, students will:

- Participate as knowledgeable, reflective, creative, and critical members of the classroom community of learners (NCTE/IRA Standard #11) who teach and are taught classic American novel(s) by one another. This will unfold as students, working in groups, negotiate what the "essential" information is from their assigned novel and determine how they will share that with the rest of the class in a way that will be interesting and memorable.
- Read (directly in their groups, but also vicariously as students learn from others) a range of print and nonprint texts to build their understanding of several American novels (NCTE/IRA Standard #1). Also, in this process, students will learn more about themselves and about other cultures, past and present, of the United States.
- Use spoken, written, and visual language to accomplish their task (NCTE/IRA Standard #12) of sharing their group's novel with the other groups so that they leave the unit with a basic understanding of several American classics. The "teaching presentations" each group shares include handouts for the class as well as posters, artifacts, and other options to facilitate visual literacy and help students to link words with tangible objects.
- Develop an understanding and respect for diversity in language use, patterns, and dialects across cultures, ethnic groups, and time periods (NCTE/IRA Standards #2 & 9) by examining a variety of novels that illuminate an array of perspectives. This will be especially apparent when students select an excerpt from their assigned text and orally interpret it for the class, therefore helping the class to "hear" the difference between authors' styles-e.g., Nathaniel Hawthorne versus Maya Angelou.
- Adjust their use of spoken, written, and visual language to effectively communicate (NCTE/IRA Standard #4) vital information about their group's assigned novel with their peers. By utilizing a variety of strategies in their presentations (oral review of the novel, accompanied by handout and/or overhead transparencies, oral interpretation, musical connections, posters, videos, and other visual aids) students will appeal to a wide range of learning styles, thus maximizing their potential to meet broad audiences.
- Employ a wide range of strategies as they write and utilize writing process elements (NCTE/IRA Standard #5). As part of this unit, students will write summaries, but also alter these to easily readable

"handouts" and "study guides" for the class. Students will also cut and edit texts in order to "tell" part of the novel's story and demonstrate the uniqueness of the author's style through an oral interpretation. As students work in groups, they will act as peer readers and editors helping one another to improve their writing.

- Apply knowledge of genre (NCTE/IRA Standard #6) to discuss, via "teaching presentations," their novel with the rest of the class. In particular, literary conventions such as character, plot, setting, conflict, and theme will be emphasized.
- Apply a wide range of strategies (including group discussion, planning and implementing presentations, writing papers, and selecting appropriate resources to help their teaching of the novel) to comprehend, interpret, and appreciate texts (NCTE/IRA Standard #3). In this process, students will draw on their prior experiences (especially as they write questions for class discussion to help make the novel accessible to all) both personally and academically, as readers of texts who look beyond individual words to conceive of holistic contexts.

Getting Started

Grouping

The teacher will determine groups (or if preferred, allow students to choose their own groups) consisting of three to six students, depending on class size and the number of novels to be studied. For example, a unit on American Classics in a class with 22 students might be divided into five groups (two groups of five students and three groups of four students). Each group would be assigned (or choose) a different American Classic (of comparable length) such as *Brave New World* (1998), *The Great Gatsby* (1996), *I Know Why the Caged Bird Sings* (1996), *The Scarlet Letter* (1992), and *To Kill a Mockingbird* (1999).

Reading Schedules

Students are to determine their own reading schedules on the day they receive their group's novel. Having a common reading schedule within each group is essential so that students are at the same point when writing dialogue journals and preparing presentations. As a guideline, students might read 20 pages per night (school nights only) or 15 pages a night (if willing to read over the weekend). Reading schedules should also take into consideration when the two "teaching presentations" occur. As a guideline, students should be about half way through their novel before the first presentation, three-fourths through the novel for the second "teaching presentation," and finished with the novel at the time the video comparison and project utilizing the Multiple Intelligences are due.

Collaborative Tasks

Performance Task for Dialogue Journals

As students read their assigned group novel, they are to write reflective entries in their dialogue journal (see *Sample Dialogue Journal*). The dialogue journal is to be handwritten on three-ring notebook paper and kept in the students' notebooks. Students will take a sheet of notebook paper and fold it vertically so that one-third of the paper is to the left of the crease and two-thirds of the paper is to the right of the crease. On the right side of the crease, the initiator of the journal entry will "reflect" on what he/she has read. "Reflections" can include (but are not limited to) the reader's thoughts, impressions, likes, dislikes, questions, and "links" to other experiences and literature. The main point is to promote active reading, questioning, and sharing. Each student is to initiate two entries (each entry is one page front and back) per week. The entries (when collected on Fridays) are only considered "complete" if they include a response from a group member. The response is written to the left of the crease (on the remaining third of the paper). "Responses" are also to be written in sentence form and may react, respond, agree, disagree, answer, or pose additional questions to the writer who initiated the entry. See *Assessment Rubric for Dialogue Journals*.

Assessment Rubric for Dialogue Journals

	Exemplary	*Acceptable*	*Needs Improvement*
Informal Response 2/3 page	Completed with evidence of active reading as demonstrated by responses to literature, initiating questions, and/ or offering extensions to other aspects of life and/or literature	Some original insight and evidence of questioning and extending, but with slight tendency to summarize	Less than 2/3 page or noticeable tendency to summarize the text, rather than responding to it
Peer Response 1/3 page	Attentive to peer's point of view; focused relevant feedback; effort to extend to the reader	Attentive and relevant overall, but 1 to 2 occurrences of wandering	Tangential, as if not paying particular attention to what was written; or simply agreeing with peer, rather than seeking to extend, challenge, or expand

Sample Dialogue Journal

Respondent (Jennifer Kerber)	*Initiator* (Kim Bertleff) *Brave New World,* pp. 1–33
You must like plenty of quotes! That's a good thing though. It shows that you are thoroughly reading and examining this book. I agree with you that something that would affect one world would eventually affect everyone. I also agree that people (or most people) make themselves the center of attention to make them feel important and needed in our society. Have you ever done that in your life? I know I have. You're right about loving nature and wanting to continue to love even more. History is mostly unpleasant, at least what we learn of it. I believe we learn the bad so we do not make the same mistakes twice. So far we do not have total peace and happiness, so I think there is a lot more for us to learn other than history. Do you agree?	So far this story is pretty weird. I'm actually understanding parts of it, but other parts are pretty confusing. The theme of Brave New World is not the advancement of science as such; it is the advancement of science as it affects human individuals. I agree with this quote because so far, I get the fact that all experimentation affects all humans not just certain people. "Later on their minds would be made to endorse the judgment of their bodies." As they grow up their minds have to develop along with their bodies." "All conditioning aims at that: making people like their inescapable social destiny." You feel, that once you've entered into a social colony, or group, it's hard to escape it due to pressure from everyone within the group. When the author says, "they're only really happy when they're standing on their heads," I feel that this is true today, because many people feel that if they're not the center of attention, they're nobody so they're not happy. Which isn't true because people will learn to love and respect you, just to be yourself. "A love of nature keeps no factories busy." I have to disagree with this quote, because if you have some love for nature you "keep going" in your life. Without that love of nature what would keep someone getting up happy every morning? I think it's a good thing to have. "But then most historical facts are unpleasant." If you think about it, this is extremely true today. Our history consists of wars and deaths, so history does tend to be unpleasant. "You can't learn a science unless you know what it's all about. " With me in biology, I agree with this. I don't understand a lot of things because I don't know what they deal with. "Ending is better than mending." I feel strongly about forgiving and forgetting. Ending is not better than mending because you lose a possibly good relationship without even trying to fix it.

Performance Task for Progress Reports

After having their books for about a week, each group is to write a Progress Report to demonstrate the group's initial understanding of its assigned novel. In addition, this ensures some advance thought and organization of ideas that should prove to be useful as groups plan their first "teaching presentation." The Progress Report (along with all other assignments except for the dialogue journals) is to be typewritten and double-spaced using the standard MLA heading with the name of the activity/assignment centered as the title. (See *Assessment Rubric for Progress Report* on page 61.) The Progress Report should include the following five items:

1. Complete bibliographic information (use *MLA Handbook for Writers of Research Papers*) for the novel
2. Summary of what students have read thus far (who, what, when, where, why, how, one to three pages)
3. Questions students are asking themselves about the novel thus far (5 to 10 questions)
4. Initial impressions/response—to promote dialogue on the novel and help to prepare students for their first "teaching presentation," each student will include his/her individual input in this section of the paper
5. Interesting points of discussion to entice the class to read the book. This should include 10 items total and at least one from each of the following categories:
   - Quotes for discussion and why these quotes were chosen
   - Hypothetical situations that relate to the novel ("what if" and "imagine" statements)
   - Relevance of the book to contemporary life
   - Thought-provoking questions that begin with the novel but then take the class beyond it

Assessment Rubric for Progress Report

	Exemplary	*Acceptable*	*Needs Improvement*
MLA Style	Complete adherence	1 to 2 errors	3 or more errors
Summary	Thoroughly addresses character, plot, setting, and conflict	Missing 1 to 2 major aspects of character, plot, setting, and/or conflict	Omissions of 3 or more aspects of character, plot, setting, and/or conflict
Mechanics/Usage	1 or no errors	2 to 3 errors	4 or more errors
Questions to demonstrate active reading reading/thinking/curiosity	8 to 10 questions	5 to 7 questions	1 to 4 questions
Items for Class Discussion (quotes, hypothetical situations, connections with contemporary life, and questions that start with the novel and go beyond it)	8 to 10 items; at least one from each of the four suggested areas	Only 5 to 7 items; or missing one of the suggested areas	4 or fewer items; or missing two or more of the suggested areas

Performance Task for Teaching Presentation

Rationale for Teaching Presentations

The bulk of this unit comes through the four presentations each group makes to teach their assigned novel to the rest of the class. Not only do students learn about the other novels through this process, but the students also increase, by having to "teach" the novel, their depth of understanding regarding the novel assigned to them.

Performance Task

Students should be about half way through their group's novel before delivering this presentation of 20 to 25 minutes. On its assigned day, each group will present to the class an overview of its novel and find a way to hold the class "accountable" for the information that was shared. Making peers accountable means they must do something (act out what the group talked about, complete an activity, take notes on a study sheet prepared by the group, respond to questions, etc.) to demonstrate that they know and understand what has just been presented. The point of the Teaching Presentations is coverage—the class goal is to learn much about five novels (or other number of novels selected by the teacher) while only reading one independently.

Because a fundamental aim is to offer the class a "cultural literacy" understanding of each group novel, the focus should be on imparting "what any reader should know" about the novel, rather than on minute or picky details. Groups are encouraged to be creative in their presenting. This may include dressing up as characters, making three-dimensional models of places, acting out scenes, composing songs, etc.

To ensure advance planning and facilitate accountability for the class, each group, on the day of its presentation, must have a two-page handout for each member of the class detailing the following:

- Characters (who's doing what in the novel; how are characters related)
- Plot (sequence of events—what's going on in the novel)
- Setting (where are we-place and time)
- Conflict (what are the issues/problems)

Each group may determine when they want to distribute the handout (before, during, or after the presentation). The handout will provide "study notes" for the class in preparation for a basic test (to be written by the teacher) on the novels that will occur after the fourth set of group presentations. See *Sample Handout for Teaching Presentations* on page 63; also note the group's creative introductory song to capture the attention of the class. The last page of the handout includes a list of questions that the group used to assist the class in being accountable for the information that had been presented. An alternative/visually creative layout for the information required in the handout may be seen in brochure form.

A second requirement for Teaching Presentation is a poster or other visual-spatial creation. Students may choose the "angle" on this creation—i.e., whether it will be used to convey information on characters, events, setting, or conflict(s). This component need not be limited to a sheet of poster board. Groups may elect to create an artifact museum (see *Artifiact Museum as Visual-Spatial Component of Teaching Presentation* (In Lieu of Poster) on page 67), make a mobile or model (see *Model as Visual-Spatial Component of Teaching Presentation* (In Lieu of Poster) on page 67), place a variety of smaller signs/maps around the room, etc.

A third required component of this presentation is the oral interpretation of literature. Here, each group is to read aloud and with emotion the most important/interesting/thought-provoking/profound excerpt from the text the group has thus far read. The purpose of this is twofold. First, students are developing their interpretive skills. Second, students are giving the class the "flavor" of the author's writing/language. This is essential because of the significant differences between authors, e.g., Hawthorne versus Huxley or Angelou. Groups are encouraged (or could be required) to "cut" and "edit" various portions of the text and divide the lines so that the class can hear multiple voices/perspectives as applicable to the interpretation. As a guideline, students would read

Text continues on page 65

Sample Handout for Teaching Presentation

Introduction

(Fred Burazer on guitar and lead vocals—class joins in on chorus)

Lyrics to the Great Gatsby

Parody to "Last Kiss"
This is the story of a single guy,
And his neighbor who is living a lie,
He holds many parties that he never attends,
He's got no family, and not many friends.

Nick has a cousin and her name is Daisy,
Her husband Tom is just a little crazy,
They have a child that's a couple years old,
And Tom has a mistress, or so we are told.

Chorus
Oh where oh where could that Gatsby be?
He's never at his own party
He's very shady and he comes and he goes,
He's got a lot of secrets that nobody knows.

One day when Nick was by himself at home,
Gatsby's butler came over all alone,
He went up to Nick and said, "If it's all right,
Gatsby would like to have you over tonight."

He was throwing a party and he thought, "Why sure,
I'll invite the guy from over next door."
Well Nick went over later and he had some fun,
He stayed until the whole party was done.

Chorus
Oh where oh where could that Gatsby be?
He's never at his own party
He's very shady and he comes and he goes,
He's got a lot of secrets that nobody knows.

Characters

Nick Carraway: He is Daisy's second cousin and he lives in a cottage in the West Egg. He lives next door to the mysterious Jay Gatsby. Nick often wonders about his neighbor who throws huge parties. Nick also went to college with Tom in New Haven.

Daisy Buchanan: She is the second cousin of Nick and Tom's wife. They live in the East Egg. Daisy is easy to upset and is affected by Tom's affair with Myrtle. She has a very difficult time dealing with the affair and depends a lot on other people.

Jordan Baker: She is a lady golfer and has been friends with Daisy for a long time. She is rich and well-bred, but she has some bad characteristics. She is reckless and dishonest, which can be witnessed in her cheating at tournaments and her bad driving. She is becoming involved with Nick.

Tom Buchanan: Tom is Daisy's husband. He lives in East Egg, the richer part of Long Island. Tom is a racist and is very commanding. He has been having an affair with a woman from New York. Her name is Myrtle. Tom is very self-centered, and tends to only apply himself to things that involve money.

Myrtle Wilson: She is Tom's mistress. She lives in New York. She is married to George Wilson, who owns a garage, even though she does not love him. She sees the glamour and the happiness of Tom's money and doesn't realize the affair is wrong.

Jay Gatsby: He is a very rich man who hosts many parties that he rarely attends. He has known Daisy for a very long time. He was Major in World War I. He claims that he attended Oxford but no one knows if that is true.

Setting The novel is set in the roaring 1920s. The characters live in the East and West Egg neighborhoods of Long Island.

Conflicts

Daisy vs. Tom: Tom is very imposing and he doesn't really appreciate his wife. He tends to yell at her a lot. Daisy knows about the affair Tom is having with Myrtle and it upsets her.

Tom vs. Myrtle: Myrtle is very jealous of Daisy and she envies the money that Tom spends on Daisy. Tom even hit Myrtle and broke her nose because she was harassing him about Daisy.

Nick vs. Jordan: Nick says that Jordan is reckless and she isn't careful when she drives. Jordan claims that it takes two reckless people to cause an accident. Nick likes Jordan but he doesn't like how she acts because he knows that she lies.

Questions for Class Accountability and Review

1. Explain how Fitzgerald uses setting to emphasize the differences between social classes.
2. In the story, Tom and Daisy are a part of the established upper class, while Gatsby is part of the class known as the nouveau riche. Decide which social group you would want to belong to and explain why.
3. Discuss what the following symbols represent in the novel:

   the valley of ashes
   the eyes of Dr. T. J. Eckleberg
   the green light at the end of Daisy's dock
   the mantle clock
   Daisy's voice "full or money"
4. Compare and contrast the characters of Tom and Gatsby.
5. Debate weather The Great Gatsby illustrates the theme of the American dream being corrupted by the desire for wealth.
6. Explain how The Great Gatsby reflects the Jazz Age.
7. Discuss what led to the downfall of Gatsby's dream.

one to three pages of the novel for this portion of the presentation. See *Assessment Rubric for Teaching Presentation #1* on page 68.

Review of Basic Components for Teaching Presentation

- Half way through novel
- 20 to 25 minutes
- Way to ensure class "accountability"
- Two-page handout to cover basics for "cultural literacy" understanding of the novel
- Poster (or other visual-spatial creation)
- Oral interpretation of literature

Be creative!

Performance Task for Project Utilizing Students' Multiple Intelligences (MI)/Presentation

Groups are to "wrap up" their investigation of their assigned novel by creating and presenting a project utilizing one (or a combination) of Gardner's Multiple Intelligences. Having students write "contracts" (to be revised/specified as necessary by the teacher, see *Sample Contract Statement and Guidelines* on page 70) a week prior to the project due date tends to enhance quality and diminish the occurrence of projects "thrown together at the last minute." The specific nature of the "revised" contract (signed by students, copied, and kept in the teacher's files until the day of presentation) will facilitate the teacher's assessment of the project.

Options for projects utilizing students' multiple intelligences are limited only by the students' imaginations. They may include paintings, drawings, dioramas, videos, songs, raps, skits, interviews, etc. The important thing for students to remember is that the focus is always on showing understanding, and especially delving into new modes of expression that unveil deeper layers of meaning.

The written component for the MI Project/Teaching Presentation should be double-spaced and typed MLA style to include the following:

- Summary of the novel (two pages)
- Group response (can be reported member by member or as a collaborative response) to the novel (one to three pages)

Written description of the group's project (how the group approached it-process-what intelligence(s) it used, and how it contributed to/enhanced the group's understanding of the novel (one to two pages).

Text continues on page 70

Alternative/Visually Creative Layout for Information Required on Handout for Teaching Presentation #1

(by Jake Gore, Shawn Michael, Mike Prince, Dan Sawyer, Ben Wittenauer)

<table>
<tr>
<td>

Left (back cover) which includes the group's selection for oral interpretation.

Right (front cover) including group members' names, novel title, author, and a quote from the text.
</td>
<td>
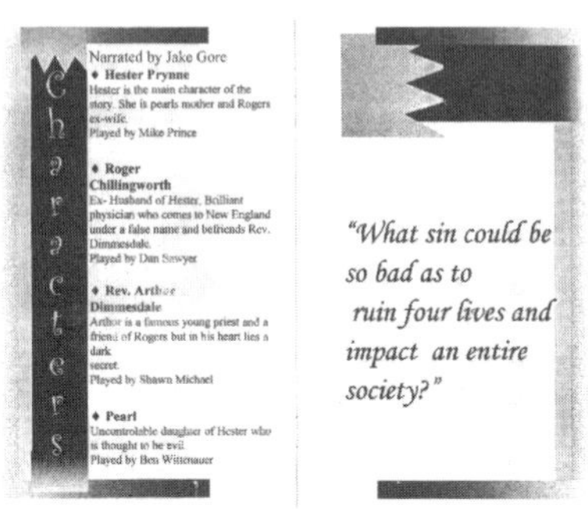

Left (page 2) lists main characters from the novel and indicates who each is "played by" as the group dresses up and tells about their character.

Right (page 3) "What sin could be so bad as to ruin four lives and impact an entire society?"
</td>
</tr>
<tr>
<td>
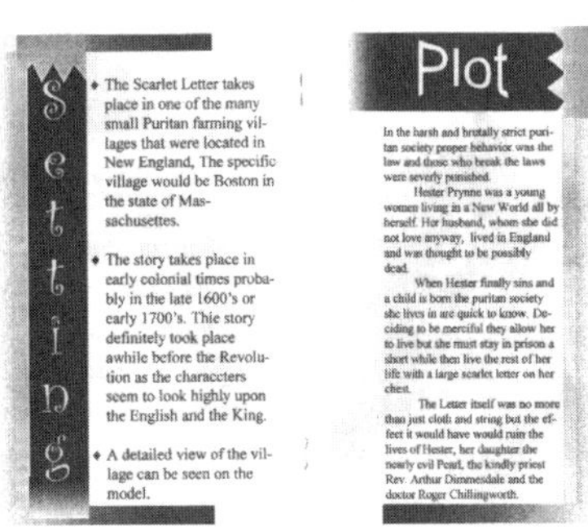

Left (page 4, center of brochure-style handout) establishes setting.

Right (page 5) reviews plot thus far.
</td>
<td>
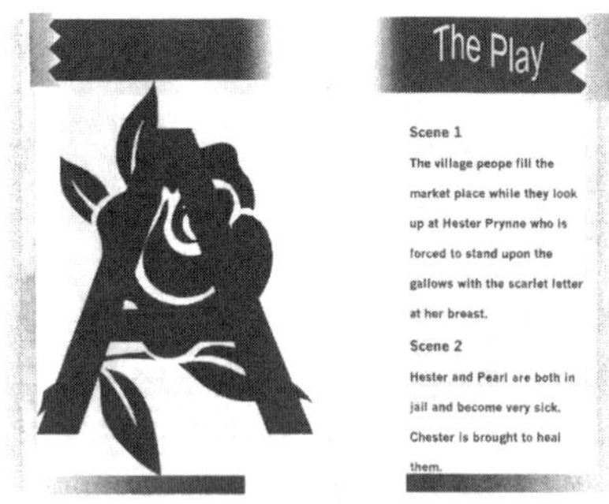

Left (page 6) Symbolic representation of the Scarlet Letter Hester Prynne wore on her dress.

Right (page 7) indicates two scenes the group will be acting out for the class as part of Teaching Presentation #1.
</td>
</tr>
</table>

Artifact Museum as Visual-Spatial Component of Teaching Presentation (In Lieu of Poster)

Jennifer Kerber, Lisa Kuzenko, Stephanie Persin, and Kim Bertleff created an "artifact museum" in addition to two "backdrop" posters for their second presentation on *Brave New World*. Each artifact represented something important from Huxley's novel and served to reiterate some of the key characters, events, and conflicts.

Model as Visual-Spatial Component of Teaching Presentation (In Lieu of Poster)

This visual was created by a group studying *The Great Gatsby*. The model depicts the opulence the group imagined would be found in Jay Gatsby's mansion. It is constructed with cardboard, covered with tissue paper, and lighted from within. The group incorporated the model into their presentation by pointing out its characteristics, such as where Nick Carraway likely met with Jay Gatsby and where the parties took place. This was accomplished as the group read excerpts from the text to support their interpretation.

Assessment Rubric for Teaching Presentation #1

(100 points possible)

Novel Title

Members

Time (20–25 minutes; 5 point deduction each 30 seconds under/over)

COVERAGE

Characters (6, 8, 10)	mentioned but not well explained	explained but without clearly establishing relationships among characters	detailed explanation with clearly established relationships among characters
Plot (6, 8, 10)	lacking in depth	detailed, but difficult for class to follow in time to take notes	detailed and easy for class to follow and take notes
Setting (6, 8, 10)	time or place (not both)	mentioned, but not well supported or established	both well established
Conflict(s) (6, 8, 10)	mentioned, but not in a deliberate fashion	explained, but without sufficient detail	explained with sufficient detail; clearly establishing relevance to rest of text

CLASS ACCOUNTABILITY (class activity, notes, quiz, game, etc.) = *DO* Method/*Originality*/Effectiveness (6, 8, 10)

ENTHUSIASM (3, 4, 5)	monotone and appearing disinterested in what you're sharing (too soft)	genuine, but lacking eye contact and personal touch to draw in the class	lively and interesting; good eye contact, volume and vocal variety
ORGANIZATION (6, 8, 10)	unclear as to who was to do what/when; appearance of planning while in the midst of presentation	basically clear, but with hesitation at times about who is to do what when; awkwardness with handouts, etc.	seamless flow of presentation
PARTICIPATION (3, 4, 5)	one or more members appearing to have no role in presentation	all participate, but not evenly distributed	balanced contributions by all members

READING OF EXCERPT FROM TEXT

	Needs Improvement	*Acceptable*	*Exceptional*
1. Vocal production (volume, tempo, articulation)	1	2	3
2. Poise (eye contact and facial expression, gesture and body)	1	2	3
3. Sense of spontaneity (newness and energy/sincerity)	1	2	3
4. Flow of speech (reflects understanding of the text)	1	2	3
5. Context (clearly establishes relevance of excerpt to rest of novel)	1	2	3
2-PAGE HANDOUT			
coverage to assist accountability (provide class with study notes)	1–2	3	4
user friendly layout & readability (proper mechanics & usage)	1	2	3
relevance to presentation/use in class	1	2	3

POSTER

	Needs Improvement	*Exceptional*
neatness	1	2
readability	1	2
use of color, pictures, other creative touches	1	2
layout (balance, composition)	1	2
relevance to presentation (clearly included)	1	2

Sample Contract Statement and Guidelines

To show our understanding of ______________________________,

(must be a specific aspect of the novel, NOT just the title of the novel)

we will ______________________________________.

(be very specific)

In the event that students turn in a statement such as:

To show our understanding of how Scout changed throughout *To Kill a Mockingbird,* we will create a scrapbook. The teacher could revise the contract by adding/specifying the following:

- The scrapbook will be 12 pages
- The scrapbook will include collages and hand-drawn pictures
- Each page of the scrapbook will include supporting quotes from the novel
- The scrapbook will include note cards and/or letters from three of Scout's acquaintances (family, friends, neighbors)
- The scrapbook will conclude with a journal entry in which Scout summarizes the three most influential events of her life in Maycomb County.

Review of Basic Components for MI Project/Teaching Presentation

- Contract negotiated between group and teacher at least one week prior to the project's due date
- The actual project utilizing multiple intelligences to be shared with the class and assessed according to contract
- Paper to include a summary of, and response to, the novel as a whole, process undertaken, and influence of final MI project on the group's understanding of the novel.

See *Assessment Rubric for Project Utilizing Multiple Intelligences/Teaching Presentation.*

Assessment Rubric for Project Utilizing Multiple Intelligences/Teaching Presentation

(100 points possible)

Novel Title ____

Members____

Contract Statement: *To show our understanding of* ____________________

we will____.

Time (not to exceed 25 minutes)

Nonadherence to contract (–10)

PROJECT (50 points) CONSIDERATIONS & TEACHER COMMENTARY

(Use of color, neatness, layout, rhythm, composure, volume, clarity, reflective depth and accuracy, etc.—as applicable. Keep in mind that this part of the rubric is general in nature because each group's project is different as negotiated by contract with teacher.)

PRESENTATION (50 points)

ROOTED IN TEXT (6, 8, 10)	not deliberate	explained, but without specific examples from text	clearly established with detailed examples from text
UNDERSTANDING (6, 8,10)	insufficient evidence of clear depth of exploration and understanding of a particular aspect of the text	basic, but not supported sufficiently to display deep understanding of "contracted" area of the text	clear depth of understanding; evident that MI project illuminated new modes of seeing the text
ENTHUSIASM (6, 8, 10)	monotone and appearing disinterested in what you're sharing (too soft)	genuine, but lacking eye contact and personal touch to draw in the class	lively and interesting; good eye contact, volume and vocal variety
ORGANIZATION (6, 8, 10)	unclear as to who was to do what/when; appearance of planning while in the midst of presentation	basically clear, but with hesitation at times about who is to do what when; awkwardness with handouts, etc.	seamless flow of presentation
PARTICIPATION (6, 8, 10)	one or more members appearing to have no role in presentation	all participate, but not evenly distributed	balanced contributions by all members

Bibliography

Angelou, M. (1996). *I know why the caged bird sings.* New York: Random House.

Gibaldi, J. (1999). *MLA handbook for writers of research papers* (5th ed.). New York: The Modern Language Association of America.

Fitzgerald, F. S. (1996). *The great gatsby.* New York: Scribner and Sons.

Hawthorne, N. (1992). *The scarlet letter.* New York: Knopf.

Huxley, A. (1998). *Brave new world.* New York: HarperCollins Perennial Library.

Lee, H. (1999). *To kill a mockingbird.* New York: HarperCollins.

7

MARKETING PROJECT— LEGO® TOYS

Margie Bush

No endeavor is worse than that which is not attempted.

Mexican Proverb

Summary

The technical context for this multimedia performance task and exercise in real-world work (business) writing begins with a letter from the famous Matt El Toy Company asking for student input on the kinds of toys the next generation of children might enjoy. Mr. El wants a complete package (toy, directions for building, advertising campaign, and so on) and employs the students as part of the Research and Development team in his company. Thus, this exercise requires students to simulate a Toy Factory by working collaboratively in teams to design, build, and market a toy. Mr. El will accept only one toy for production from this entire group, so each team will compete against the other for the best product and ad campaign. The marketing project will culminate with a presentation at a gala open house in front of other students, faculty, and invited guests from the community.

A more detailed account of this project may be found in the December 1995 issue of the English Journal, in an article entitled "Promoting Active Learning and Collaboration Through a Marketing Project," by Jacqueline Glasgow and Margie Bush.

NCTE/IRA Standards

First, this activity requires students to read a wide range of print texts to build an understanding of marketing strategies; the writing of proposals, feasibility studies, and directions; advertising campaigns, and so on. In performing these research tasks, students will gain knowledge of themselves (especially as consumers), and of the cultures of the United States and the world in order to acquire new information, to respond to the needs and demands of society and the workplace; and for personal fulfillment (NCTE/IRA Standard #1). Therefore, this activity requires students to apply a wide range of strategies to comprehend, interpret, evaluate, and appreciate texts by drawing on their own experience, their interactions with other readers and writers, their knowledge of word meaning and of other texts, their word identification strategies, and their understanding of textual features (NCTE/IRA Standard #3). In writing their proposals/feasibility studies, and in designing their toy and advertising campaigns, students will adjust their use of spoken, written, and visual language, e.g., conventions, style, and vocabulary to communicate effectively with a variety of audiences and for different purposes (NCTE/IRA Standard #4). To produce their product, students will use a variety of technological and information resources, such as digital/video cameras, scanners, and desktop publishing programs, to gather and synthesize information and to communicate knowledge (NCTE/IRA Standard #8). And, finally, students will use spoken, written, and visual language to accomplish their own purposes, e.g., for learning, enjoyment, persuasion, and the exchange of information (NCTE/IRA Standard #12). In the final presentation, students will perform their commercials, present their product, and compete for the privilege of producing their toy.

Overview of Authentic Learning

Students, regardless of age, will become immediate and enthusiastic partners in this real-world exercise. They will begin by studying from various technical writing text books that the teacher can obtain at the library or local technical college if they are not already in the school. The study of technical writing is especially useful in adding to the students' repertoire of writing and speaking skills. If community resources are available, the teacher or students should contact people in advertising, radio, and TV for advice and even samples of advertising used in each venue. For the class, the students will need an Inquiry-Based Letter from the Matt El Toy Company, an Instruction Sheet, a Time Line of due dates, and an Assessment Rubric when they begin. The teacher will want to read over the entire project, modify the project to fit the students and class, and generate the necessary assignment sheets. The teacher should also be prepared to amend the time lines as students progress through the project.

To close out the assignment, students might invite teachers, community members, and students to a "viewing" of the completed toys, during which each group will conduct a formal presentation. The audience might take on the role of a toy company and select the toys and marketing campaigns that should be accepted by the company for production. This provides a real-life audience and gives opportunity for feedback.

Teacher Supplies

- Assignment Sheets
- Assessment Rubric
- Information about technical writing, either in the form of books, "how to" sheets, or brochures that explain such technical writing elements as:
  - proposals
  - feasibility reports
  - introduction
  - description of mechanism
  - instructions
  - troubleshooting guides
  - do's and don'ts
  - graphic views
  - advertising campaigns (radio/TV commercials and print ads)
  - warranties, guarantees
- Most students are familiar with marketing strategies as consumers. They might want to generate complimentary material to go with their toy. This might include such things as:
  - fan club info
  - certificates
  - complimentary toys
  - accessories
- Work space and shelf space to keep the toys once assembled
- Sample manuals, instruction booklets, brochures

Sample advertisements, TV/radio ads and print media ads (community resources will be helpful here)

- Time: if this is to be a true group project, the teacher will need to provide in-class time for students to work together.

Student Supplies

- Legos®, Tinker Toys®, or other connecting blocks
- Plastic bag to hold excess pieces
- Shoe box or box to hold toy once it's assembled
- Binder for the manual once it is assembled

- Optional things students might use-camera/film, sheet protectors
- Decorated box-as it would appear on the store shelf

Planned Performance Tasks

- Present students with a letter inviting them to participate in this project. You may use the one provided in this assessment or modify it to suit your needs. See *Sample Inquiry-Based Letter* on page 77.
- Divide students into groups and distribute or ask students to bring in Tinker Toys, Legos, or other individual building blocks. Ask them to begin creating a toy. After they reach consensus on the construction of the model or prototype, ask each group to develop a project proposal. The proposal can be written collaboratively and should include details about the project: objectives and need for the toy, methodology, schedule, price, and qualifications of personnel. See *Marketing Package Requirements* on page 76 and *Assessment Rubric for Project Proposal* on page 78.
- Once the project proposals are approved, the students then design and compose a Marketing Instruction Manual. The components of the manual include: an historical introduction, equipment list, instructions for assembly (side, front, rear views), and a troubleshooting guide. See *Assessment Rubric for Marketing Instruction Manual* on page 79.

Marketing Package Requirements

Binder
Cover
Cover Letter
Table of Contents
Introduction
Proposal
Feasibility Study
Parts with Diagrams or Pictures
Description of Mechanism
Exploded Views
Front, Rear, Side Views
Assembly Instructions
Troubleshooting Guides
Do's and Don'ts
Press Release
Copy for TV/Radio Ads (Storyboard)
Copy for Print Ads
Warranties, Guarantees
Miscellaneous Material

Sample Inquiry-Based Letter

MATT EL TOY COMPANY, INC.
2250 Avenue of the Americas
New York, New York 10002
212–555–5555

English 11 Advanced Students
Shawnee High School
3333 Zurmehly Rd
Lima OH 45806

5 December 2000

Dear Students,

On behalf of Matt El, Inc., I am inviting you and your classmates to participate in a wonderful opportunity to put to work your creative writing and speaking abilities. The Matt El Toy Company has noticed a considerable lack of inventive, educational, and creative toys in today's market. We believe that you, as teenagers, possess a unique perspective on both the consumer market and on the dreams of kids.

We would like you to design your own toy and marketing campaign and compete for the right to have it manufactured by our toy company.

Your teacher will share with you some of the requirements of this opportunity and tell you how you will proceed.

Sincerely,

Matt El

Matt El
CEO, Matt El Toy Company

Assessment Rubric for Project Proposal

	Exemplary (4)	*Proficient (3)*	*Apprentice (2)*	*Novice (1)*	*Score*
Objectives	Clearly stated and complete description of the purposes and the need to build the toy	The proposal provides adequate description of purpose and need to build the toy	The proposal provides some of the purposes, but misses key points in the purpose and/or need	The proposal does not provide the purposes or need to build the toy	__× 4 = __ out of 16
Key Features	The proposal details both key and hidden features of the toy and explains how they serve several purposes	The proposal details the key features of the toy and explains the purposes they serve	The proposal neglects some features of the toy or the purposes they serve	The proposal does not detail the features of the toy or the purposes they serve	__× 6 = __ out of 36
Schedule	Schedule is ambitious given the time frame	Schedule is realistic for the time frame	Schedule is not realistic for the given time frame	Schedule doesn't work for the given time frame	__× 2 = __ out of 8
Price	Price is realistic based on evidence of research	Price is in the ball park, but needs more evidence of research	Price is too high or too low; research not cited	Price is not reasonable and there is no evidence of research	__× 3 = __ out of 12
Qualifications of Personnel	Creative description of personnel	Adequate description of personnel	Insufficient description of personnel	No description of personnel	__× 3 = __ out of 12
Conventions	Correct sentence structure, grammar, punctuation and spelling	Some errors in sentence structure, grammar, punctuation and/or spelling	Distracting errors in sentence structure, grammar, punctuation and/or spelling	Too many errors in sentence structure, grammar, punctuation and/or spelling	__× 4 = __ out of 16
			Total Score (100 possible)		
			Grade	A = B = C = D = F =	

Assessment Rubric for Marketing Instruction Manual

	Exemplary (4)	*Proficient (3)*	*Apprentice (2)*	*Novice (1)*	*Score*
Introduction	Provides clear background information context, purpose, organization, audience for instructions	Provides adequate background information context, purpose, organization, audience for instructions	Provides inadequate background information context, purpose, organization, audience for instructions	Does not provide background information context, purpose, organization, and/or audience for instructions	__ x 4 = __ out of 16
List of Materials and Tools	Elaborate list	Adequate list	Inadequate list	No list	__ x 3 = __ out of 12
Instructions for Assembly	Complete step by step instructions written in active voice; clear explanations with precise details	Step by step instructions written in active voice; good explanations with significant details	Missing some steps or insufficient explanations and details	Many steps missing; incorrect explanations; and/or inadequate details	__× 6 = __ out of 36
Diagrams	Clear, effective diagrams showing front, rear, side views; exploded views to facilitate assembly	Good, clear diagrams showing front, rear, side views; exploded views to facilitate assembly	Confusing diagrams showing front, rear, side views; exploded views to make assembly difficult	Diagrams are incomplete and make assembly impossible	__× 5 = __ out of 20
Trouble-shooting Guide	Articulates what to do if something goes wrong with the assembly or product	Good description of what to do if something goes wrong with the assembly or product	Inadequate description of what to do if something goes wrong with the assembly or product	No description of what to do if something goes wrong with the assembly or product	__× 3 = __ out of 12
Do's and Don'ts	Provides audience with a clear list of warnings to ensure safety	Provides audience with a good list of warnings to ensure safety	Doesn't provide audience with a good list of warnings to ensure safety	No list of warnings to ensure safety	__× 4 = __ out of 16
Warranties/ Guarantees	Creative description of warranties and guarantees	Adequate description of warranties and guarantees	Inadequate description of warranties and guarantees	No warranties and/or guarantees	__× 3 =__ out of 12
Conventions	Correct sentence structure, grammar, punctuation and spelling	Some errors in sentence structure, grammar, punctuation and/or spelling	Distracting errors in sentence structure, grammar, punctuation and/or spelling	Too many errors in sentence structure, grammar, punctuation and/or spelling	__× 4 = __ out of 16
			Total Points		
			Grade	A = B = C = D = F =	

- When manuals are completed, students should devise a multimedia advertising campaign. They can create radio and television commercials as well as print ads. See *Student Radio Script* on page 81 and *Student TV Advertisement* on page 82.
- Once the advertising campaign is organized, ask students to prepare a multimedia presentation of their toy and marketing strategy. Set a place, date, and time for the presentations. Invite parents, school personnel, and community members to attend. Invite the audience to assess the projects and decide which ones should be developed further. See, on page 83, *Assessment Rubric for Oral Presentation*, adapted from the *Massachusetts Speaking Assessment Criteria.*

Bibliography

Chicago Board of Education (2000). *Massachusetts speaking assessment criteria.* Assessment of Basic Skills, Speaking Assessment Rating Guide [Online]. Available: http://intranet.cps.kiz.il.us/Assessments/Ideas_and_Rubrics/Intro_Scoring/intro_scoring.html.

Glasgow, J., & Bush, M. (1995, December). Promoting active learning and collaboration through a marketing project. *English Journal*, 84, 32–37.

Student Radio Script

WSHS-Radio 92.1FM
3333 Zurmehly Rd
Lima OH 45806
419–555–5555

General Manager: Donald J. Wade

RADIO SCRIPT
(:60 second ad)

TITLE:	CLIENT:
	DATE:
WRITER:	TIME:
	REMARKS:

Background Party Scene: Cue "Happy Birthday" Music (instrumental)

Father:	Here, son, open this one. It's from us. (*Cue unwrapping sound effect*)
B Boy:	(*Sarcastically*) Oh, great, another toy spaceship.
Child:	This one's from me! Open it, Open IT!!!! (*Cue unwrapping sound effect*)
B Boy:	Thanks, Bobby! (*With disappointment*) Oh, another toy truck. (*Cue tossing onto pile sound effect*)
B Boy:	Hey, it's Joey!
Joey:	Here's my present. Open it up. (*Cue unwrapping sound effect*)
B Boy:	(*Excited*) An InterGalactic FunStation!!! There's so much to do! It's got a battlestation with an escape route, Intergalactic swings, the tire powered energy station, and the InterGalactic soldiers come right with it! Wow! It's great.
Father:	Joey, this must have cost you a fortune!
Joey:	Nah, it was just $39.99
B Boy:	This sure beats a toy truck!
Joey:	And, it's got tons of room to add whatever you want… AND there's Inter-Galactic Fun station T-shirts, hats, and even a CD!
B Boy:	Thanks, Joey, this is the best birthday present ever!
Joey:	The InterGalactic Fun Station: The Choice is Yours!

Student TV Advertisement

SHS TV-35
An affiliate of NBC
3255 Zurmehly Rd
Lima OH 45806
419–555–5551

TV Advertisement
(:30 second TV Spot for Prime Time Networks)

TITLE:	CLIENT:
	DATE:
WRITER:	TIME:
	REMARKS:

(Story Within Box)

(Open with a view of competitor's or similar product, i.e., vehicle)

Announcer: Tired of the same old boooorrrrring trucks and cars? (vehicle is smashed by sledge hammer with shattering sound effects)

(Shot of competitor's spaceship)

Announcer: Is there a stack of spaceships in your room that you never play with? (ship is set ablaze by a blowtorch)
(Shot of open sky)

Announcer: Sick of your friends buying you worthless junk that you'll never use? (competitor's product flies across the screen and is blown to bits by a shotgun blast)

(Cue InterGalactic Fun Station Theme)

(Cue rotating product shot)

Announcer: No need to worry, now that there's the new InterGalactic Fun Station, the new and totally different toy that you won't want to put down! Ready for battle or just for a good time, the InterGalactic Fun Station comes equipped with a battle station, swing, wheel run, and rotating platform for shooting those bad guys, and is available at your local toy store for the low, low price of $39.99. Tell your parents to get you one today! The InterGalactic Fun Station: The Choice is Yours!

Assessment Rubric for Oral Presentation

	Exemplary (4)	*Proficient (3)*	*Developing (2)*	*Beginning (1)*	*Score*
Content	The content is superior in meeting the requirements of the assignment	The speaker provides enough content to meet the requirements of the assignment	The speaker does not provide enough content to meet the requirements of the assignment	The speaker says practically nothing	__× 9 = (36 points)
Organization	The message is well-organized; the speaker provided transitions and clue words to indicate ideas, details, opinions	The message is organized and the sequence and relationships among ideas is clear	The organization of the message is mixed up; the organization appears to be random or rambling	So disorganized one cannot understand most of the message	__× 6 = (24 points)
Speaking Skills	The speaker delivers the message in a lively, enthusiastic fashion	The pronunciation and enunciation are clear; appropriate volume and speaking rate	The audience must work to understand the message; the volume is too low and the speaking rate too fast	The pronunciation and enunciation are so unclear that one cannot understand most of the message	__× 5 = (20 points)
Language	The speaker makes very few grammatical mistakes; the speaker uses language in effective ways to emphasize or enhance the meaning of the message	The speaker makes few grammatical mistakes; the speaker uses language that is appropriate for the task	The speaker makes many grammatical mistakes; the speaker uses very simplistic, bland language	The grammar and vocabulary are so poor that one cannot understand most of the message	__× 5 = (20 points)

Total Points (100 possible)

Grade Scale A =
B =
C =
D =

8

FAMILY HISTORY MULTIGENRE PROJECTS

Nanci Bush

Life is not a problem to be solved, but a reality to be experienced.

Soren Kirkegaard

Overview of Authentic Learning

In this assessment, students connect their own ancestral heritage to historical events and influences. Much of the research for this project is completed through personal interviews of family members, supplemented by textual sources. Students write four pieces of historical fiction, which are presented in both written and oral forms. By sharing individual stories within the classroom, students get a unique perspective on the rich diversity of the United States. This unit easily fulfills research requirements in English courses and could easily be adapted for history or cross-disciplinary courses.

The Family History Multigenre Project is a student-centered approach to research and learning. As students research their families' immigration experience, their engagement is high, and meaningful learning occurs. This research requires that students make interdisciplinary connections and use higher-level thinking skills to gather information, interview relatives, compare family stories to historical events, and report them in nontraditional formats. Instead of the traditional research paper format, students report their research in a series of creative pieces that tell a story. See Romano's Blending Genre, Altering Style: Writing Multigenre Papers for additional information about this type of research project. By asking students to compose creative pieces, they learn to fuse their academic and personal voices. Students nurture their multiple intelligences through active research strategies, and creative oral and written presentations of their research.

NCTE/IRA Standards

During this five-week unit, students will:

- Research connections between student's ancestors and historical events. In doing so, students will read a wide range of print and nonprint texts to build an understanding of texts, of themselves, and of the cultures of the U.S. and the world; to acquire new information; to respond to the needs and demands of society and the workplace; and for personal fulfillment (NCTE/IRA Standard #1); they will conduct research on issues and interests by generating ideas and questions, and by posing problems. They gather, evaluate, and synthesize data from a variety of sources (e.g., print and nonprint texts, artifacts, people) to communicate their discoveries in ways that suit their purpose and audience (NCTE/IRA Standard #7); and they will develop an understanding of and respect for diversity in language use, patterns, and dialects across cultures, ethnic groups, geographic regions, and social roles. (NCTE/IRA Standard #9).
- Utilize factual information as the basis for fictionalized stories. In doing so, students will apply a wide range of strategies to comprehend, interpret, evaluate, and appreciate texts. They draw on their prior experience, their interactions with other readers and writers, and their knowledge of word meaning and of other texts (NCTE/IRA Standard #3), and students whose first language is not English can make use of their first language to develop competency in the English language arts and to develop understanding of content across the curriculum (NCTE/IRA Standard #10).
- Practice writing in a variety of creative genres. In doing so, students will employ a wide range of strategies as they write and use different writing process elements appropriately to communicate with different audiences for a variety of purposes (NCTE/IRA Standard #5).
- Consider the topic from various vantage points. In doing so, students will use a variety of technological and informational resources (e.g., libraries, databases, computer networks, video) to gather and synthesize information and to create and communicate knowledge (NCTE/IRA Standard #8).
- Unify the genres through a structure or a *repetend* (repeated phrase). In doing so, students will employ a wide range of strategies as they write and use different writing process elements appropriately to communicate with different audiences for a variety of purposes (NCTE/IRA Standard #5).

- Utilize writing processes to develop a quality product. In doing so, students will adjust their use of spoken, written, and visual language to communicate effectively with a variety of audiences and for different purposes (NCTE/IRA Standard #4); they will employ a wide range of strategies as they write and use different writing process elements appropriately to communicate with different audiences for a variety of purposes (NCTE/IRA Standard #5); and they will apply knowledge of language structure, language conventions, media techniques, figurative language, and genre to create, critique, and discuss print and nonprint texts (NCTE/IRA Standard #6).
- Share ideas with classmates, informally and formally, through a peer review process. In doing so, students will participate as knowledgeable, reflective, creative, and critical members of a variety of literacy communities (NCTE/IRA Standard #11).
- Present research findings to the class. In doing so, students will adjust their use of spoken, written, and visual language to communicate effectively with a variety of audiences and for different purposes (NCTE/IRA Standard #4), and students will use spoken, written, and visual language to accomplish their own purposes (NCTE/IRA Standard #12).

Performance Tasks for the Family History Multigenre Project

This project really combines two related performance tasks: a multigenre paper written about the student's family/ancestors and a presentation of the information to the class.

- As a way to generate enthusiasm for the project, I begin by asking students what they know about their family's immigration experiences. (See *Student Handout* on page 88.) Typically the responses range from students who immigrated to the United States within the past ten years to students who are related to someone from America's colonial period or to Native Americans. Then I show a clip from the video, Ellis Island (1997), and we discuss the major influx of immigration that occurred between 1892–1924 when over 22 million passengers and ships' crews came through Ellis Island and the Port of New York. Students can research passenger records from the ships that brought the immigrants and build a family scrapbook at the American Family Immigrant History Center Web site: http://ellisisland.org. If time permits, students can participate in *Gateway: A Simulation of American Immigration History* by Interact. In this simulation, students are assigned specific identities in 1900, as Greeks, Italians, Russians, Scandinavians, Germans, or Poles. They assume roles of either processors at Ellis Island or immigrants who are entering America, "The Promised Land." They receive specific role-play information on age, health, character, vocation and political beliefs of "their" immigrant. Then the classroom is organized as the interior halls of Ellis Island were, enabling you to "process" immigrants through Background, Character, Vocation, and Health stations. The "immigrants" gain admission to the U.S. and take the loyalty oath or are deported for health or character reasons. Through these activities, students can procure an immigrant story (hopefully their own, but if not, a fictional one) to provide consistency throughout the student's research.

Student Handout

Give me your tired,
your poor,
your huddled masses
yearning to breathe free,
the wretched refuse of your teeming shore.
Send these, the homeless,
tempest-tost to me.
I lift my lamp beside
the golden door.

Emma Lazarus 1883

This is the inscription on the Statue of Liberty.

The United States is a country built on immigration. What is your family's experience? When did your earliest family members first arrive? What about other family members? Through which port(s) did they come? How did they survive as immigrants? What challenges did they face? What obstacles were they able to overcome?

- After generating ideas through freewriting, journal entries, and discussion, I distribute a handout to students detailing the instructions for the multigenre research project and due dates. (See *Student Assignment Sheet*.) I also share student samples with the class so that they can better envision their end products. I assign this project in mid-November although the final isn't due until May (although this assessment could be adapted to a more concentrated unit completed in a grading period of five or six weeks). The longer time frame allows students to take advantage of family gatherings throughout the year.

Student Assignment Sheet

Family History/Multigenre Paper & Presentation

Overview

Over the next five months you will need to contact and interview relatives, compare your family stories to history, and write four pieces that narrate your ancestors' roles throughout history. Beginning in January, you will submit a monthly journal detailing your progress and eventually including your drafts. After your final papers are submitted, we will begin an oral sharing of the stories you've garnered from family members.

A multigenre research paper relies on research to learn about a subject, but then it is presented in a nontraditional format. The format relies on various genres that are created by the writer to convey a more personal side of the subject than could be expressed by a traditional paper. You will convey factual details in a more creative way.

A genre is a form of writing. Genres often used in these papers include: short stories, poems, plays, essays, newspaper articles, letters, journals, vignettes, interview conversations, etc. Your paper should include at least four different historical events, each recorded in a different genre.

Responsibilities

First, begin interviewing family members for the most engaging stories that have happened to members of your family. Try to determine when and how your family immigrated to the United States. Other stories should reveal what it means to be an American at a certain time. Ideally, the stories will tie to a "historical event." Sometimes, you may be able to talk to primary sources (the actual person who experienced the event); otherwise, you may talk to secondary sources. As you research you will need to keep track of your sources for a bibliography page, just like in a traditional research paper—except your sources are mostly relatives, not books. (Follow the MLA format in *Writers INC.*)

Second, to help you remember the project and to allow us to communicate about your work, you will need to complete a two to three page journal entry each month discussing and demonstrating your progress. Record all of your thoughts, ideas, tips, notes from sources, quotes, and beginning stories about your family. At first, you might use this space for initial planning; then you should use this space to draft your genres.

Third, you will begin finalizing the pieces that will compose your final paper. The genres that you choose are up to you and should blend well with your subject.

The best papers will be cohesive (i.e., there will be some unity holding them together). Perhaps you will establish a structure in which this grouping of materials could logically be found. Or, you could connect your pieces with a repetend (a group of words, a phrase, or an image that is repeated exactly, unexpectedly), another useful device for achieving unity.

After you complete the final paper, create a notespage commenting on the details you've added to the piece and those that are factual and reflecting on your family's role in this historical event. A bibliography is also required.

Due Dates

Journal entries (2–3 pages) are due ______________________

Final papers are due ______________

Presentations begin ______________

Minimum Requirements

4 genres, 4 historical events, 4 pages, a structure or a repetend to connect the genres, a learning log throughout the early phases, and an oral presentation after the written work is complete.

- At some point, it is important to ask students to brainstorm all the genres and modes of expression they know. They should be able to list at least 50 genres from their experiences and culture that might include: liner notes of CD's, music video scripts, types of pieces in the newspaper, magazines, etc. Examine multigenre novels or fiction in new formats such as those found in *Young Adult Multigenre Novels or Fiction in New Formats*. Ask students to make a list of genres used by these authors. Read some of them aloud and discuss what the author was able to do in that genre or style. As you acquire multigenre papers from year to year, you might ask students to "perform" them by giving students a piece to read. Then you can have the class do a similar analysis of those texts.

Young Adult Multigenre Novels or Fiction in New Formats

Avi. (1991). *Nothing but the truth.* New York: Avon Flare Book.

Bode, J., & Mack, S. (1998). *Hard time: A real life look at juvenile crime and violence.* New York: Laurel Leaf Books.

Chbosky, S. (1999). *The perks of being a wallflower.* MTV/Pocket Books.

Cole, B. (1997). *The facts speak for themselves.* New York: Front Street.

Cormier, R. (1979). *I am the cheese.* New York: Knopf.

Danziger, P., & Martin, A. (1998). *P.S. longer letter later.* New York: Scholastic.

Deem, J. (1994). *3 Nbs of Julian Drew.* New York: Houghton Mifflin.

Draper, S. (1994). *Tears of a tiger.* New York: Atheneum.

Fleischman, P. (1999). *Mind's eye.* New York: Holt.

Fleischman, P. (1999). *Seedfolks.* New York: HarperTrophy.

Marsden, J. (1991). *Letters from the inside.* New York: Laurel Leaf Books.

Sebestyen, O. (1988). *The girl in the box.* New York: Dell Laurel Leaf.

Spiegelman, A. (1997). *The complete Maus: A survival's tale.* New York: Pantheon.

Thomas, R. (1996). *Rats saw God.* New York: Simon & Schuster.

Wolff, V. E. (1998). *Bat 6: A novel.* New York: Scholastic.

Wolff, V. E. (1997). *Make lemonade.* New York: Holt.

Woodson, J. (1995). *From the notebooks of Melanin Sun.* New York: Scholastic.

Yolen, J., & Coville, B. (1998). *Armageddon summer.* Harcourt Brace.

- During second semester, I begin having one due date per month to help students refocus on the project. I find this generates better-quality work because students are not able to procrastinate, and because they have more wait time before revision of the final pieces. On the checkpoint dates, I ask students to share examples of their writing with the class. This seems to generate ideas and models for other students to emulate. I ask for students to share examples of poetry, creative use of graphics, news articles, letters, dialogues, stories, and other genres they have generated. At this point, I ask them to write an "End Note" for the piece they prepared for this day. This is not the traditional MLA End Note. End Notes for the multigenre paper are the place that students distinguish between what is factual and what has been fictionalized. They might identify the source of their inspiration for the piece or reveal information about the writing processes and research information. Students need to include an End Note for each of their pieces. See *Student Poem and End Note* written by Beth Divis, Solon High School, May 2000.

Student Poem and End Note

John Domsic, great grandfather

It seems to be
a part of my family's history,
that my great grandfather
(craving wine that was illegal to consume in 1919)
set up a small business, in the basement of 125, East 63rd
Cleveland, OH.

The same basement where my grandpa and his brother
were caught smoking up the chimney
(they were promptly taken to the drug store where
expensive cigars were bought for them).

He'd order grapes from Ashtabula county,
as my grandpa would say, itching the tip of his nose,
slyly looking both ways, that old Croatian man, who altered his last name to sound more American, would slide the crates
down wooden slats, into the basement coal bin
(he was careful to send for the grapes in late summer before the coal was delivered).

Decades later, my mom remembers walking into that basement
as a little grand daughter:
it was dark like memories, stank like vinegar;
was wooded and damp.
Barrels bigger than a six-year-old girl, lay on their sides
and made the walls look convex.

That old Croatian man was used to a glass with dinner,
how wrong to outlaw a part of his tradition.
So dark like the cave,
he'd pack his crates into rows that lined the coal bin.

Today my grandpa offers to bring the wine
to our Thanksgiving and Christmas celebrations.
The boxed wine he carries, tucked under his arm—
The man who remembers his hat before he goes out.

End Note—

My grandpa sparked my interest on prohibition over stuffed cabbage a few weeks ago. My parents and I were eating dinner with him when the subject came up. He talked about how his dad began making wine in the basement after prohibition. I was interested in this because I had never heard anyone speak of my great grandfather Domsic before my grandpa did. I asked him to elaborate on his father's wine making and then conducted an informal interview with my grandpa.

My grandpa told me that his father ordered g rapes from Ashtabula county. Since he did slide them into the basement via the coal shoot, I assumed that he was careful to send for the grapes in late summer. I also assumed that this man, who was ultimately breaking the law in making wine, had an element of secrecy and slyness about him.

My mom contributed her memories of the basement. She remembered it to be a dark, wooded place with a very low ceiling.

I connected my grandpa to his law-defying father in the last stanza. I thought this stanza would show how there is a level of tradition in my family. Because wine was such an important cultural part of his childhood, my grandpa offers to bring it to our family celebrations today. I can't imagine how strange it must have been for so many immigrants to give up alcohol in accordance with a governmental ruling. It had been a part of their lives in the old country and they expected to follow their traditions in America.

The last line of the poem is a solid fact. My grandpa always looks his best and never forgets his hat before going out. This connects to the dignity alluded to in the second to last stanza.

- Several weeks before the final paper is due, I review the assignment criteria and the assessment rubric with the students. This builds in an opportunity for student questions and teacher refocusing of the project. As they begin finalizing their pieces, we talk about how to organize them so that they tell a coherent story. Students need to think about how they will make transitions between the pieces so there is unity and coherence to the multigenre paper. Romano writes about a rhetorical device called a *repetend,* which is "the unexpected repetition of a word, phrase, sentence, or passage" (2000, p. 154). Ask students to bring in examples of repetition, repetend, and recurring images from literature, videos, and film. Romano suggests asking students to invent words, phrases, sentences, or images they could repeat in their papers (p. 164). Students find creative repetends to connect their pieces with graphics, photos, quotes, facts, statistics, rules, historical events, letters, comics, etc. Some recent student repetends include: "America the beautiful," "the land of opportunity," "breaking away from tradition," "never again," and "think for yourself". Such repetends truly build in sophistication in the hands of a skilled writer. For instance, in the paper with "the land of opportunity" repetend, the author had an ancestor ask the question, "But I thought this was the land of opportunity?" While some students italicize these lines for impact, some students hide them in the text for the readers' delight. The last written part of the project is constructing the bibliography. Students must accurately document all research sources using MLA style of documentation.

- On another day, we discuss more detailed plans for the presentation, assign presentation dates to individuals, and review the assessment rubric for the presentation. By using the rubric as a teaching tool, I am able to clearly define my expectations for the project, and student ambition and success tends to be higher. See *Assessment Rubric for Multigenre Papers* on page 94 and *Assessment Rubric for Oral Presentations* on page 95.

Bibliography

Interact Middle School Catalogue (2001). *Gateway: A simulation of American immigration history.* Carlsbad, CA: author. (Available: www.interact-simulations.com/)

Romano, T. (2000). *Blending genre, altering style: Writing multigenre papers.* Portsmouth, NH: Boynton/Cook.

Sebranek, P., Meyer, V., & Kemper, D. (1992). *Writers INC: A guide to writing, thinking & learning.* Burlington, WI: Write Source Educational Publishing House.

(1997). *Ellis Island* [Video]. A&E Entertainment, HistoryChannel.com.

Assessment Rubric for Multigenre Papers

	Exemplary	*Admirable*	*Adequate*	*Minimal*
Depth of Content	Contains four or more thoughtful and thorough genres	Contains four thoughtful genres	Contains four genres	A lack of effort is evident. Genres are brief or contain logical inconsistencies
Connections to History	Demonstrates an understanding of what ancestors must have experienced during various historical events	Accurately depicts historical ties	Connects to four historical time periods	References to historical events may seem superficial
Quality of Writing	Demonstrates impressive quality of writing and few grammatical errors	Demonstrates good writing ability with only minor grammatical errors	Writing completes its purpose, but without maintaining a lasting impression on the reader; grammatical errors may be present	Writing reveals lack of effort or lack of language skills; grammatical errors may make the paper difficult to read
Unity	A unique repetend or structure unites the piece	A repetend or structure unites the piece	A repetend or structure unites the piece but it may seem contrived or trite	Pieces may seem disjointed
Notespage	Includes a notespage that reflects on the actual event and the liberties taken within each genre and the writing process	Includes a notespage that reflects on the actual event and the liberties taken within each genre	Includes a notespage that mentions the actual event and the liberties taken within each genre	Notespage only shallowly acknowledges source of information
Bibliography	Accurately documents sources using MLA format	Accurately documents sources using MLA format	Documents sources using MLA format	Documents sources

Assessment Rubric for Oral Presentations

	Exemplary	*Admirable*	*Adequate*	*Minimal*
Content	Presents memorable stories that evoke an emotional response in viewers. Tie to history is integrated	Presents stories which interest audience. Tie to history is evident	Presents stories which generally interest audience. Tie to history seems limited	Presents stories which tie to history in a superficial manner
Organization	Presentation includes a strong introduction, a logical organization, and an effective conclusion	Presentation includes a logical organization with a good introduction and conclusion	Presentation includes a logical organization, though introduction and conclusion may seem formulaic	A lack of preparation seems evident. Materials and ideas may seem unorganized
Visual Effects	Student may sing a song, play an instrument, or share a self-created video or display other creative self-expression	Student may share a related video clip or artifacts that demonstrate a concept from the presentation	Student provides a poster or family pictures for audience to relate to	A lack of effort in finding a related visual aide is evident
Eye Contact	Student establishes good eye contact with audience and makes audience members feel as connected	During most of the presentation, student maintains eye contact with audience	During part of the presentation student maintains eye contact with audience	Student establishes minimum eye contact with audience
Vocal Projection	Student enunciates clearly, uses expression, and is clearly heard by the audience, regardless of "outside noise"	Student enunciates clearly and audibly and makes several attempts to integrate expression	Student speaks clearly, but with little expression	Student does not enunciate and is difficult to hear
Body Language	Student uses gestures and mannerisms to enhance the delivery and does so naturally and effectively	Student uses gestures and mannerisms to enhance the delivery	Student seems stiff, gestures may seem forced	Student uses few or inappropriate mannerisms that detract from the presentation
Appearance	Student may dress in costume or other dress attire	Student dresses appropriately for a school presentation	Student dresses neatly, avoiding jeans and tennis shoes	No effort to improve daily appearance is obvious

9

PROMOTIONAL MAGAZINES FOR LOCAL BUSINESSES

Margie Bush

The smallest act of kindness is worth more than the grandest intention.

Anonymous

Overview of Authentic Learning

"Big Business wants to hire employees that understand their operation and environment, as well as the current and emerging issues that they face. Big Business also wants to hire people that know how to work together productively in teams," says Dennis R. Poux, Coordinator of Training and Organizational Development, Thomas Steel Strip Corporation, Warren, Ohio. To this end, the following simulated business experience provides an opportunity for teams of high school students to choose a local business and study the operation.

Through on-site visitation, interview and research, students develop a promotional magazine for that business to sell their product, service, or equipment. This project demonstrates how the community and workplace can be invaluable extensions of the classroom. Students study authentic models of technical writing through hands-on activities that include reading, writing, listening, speaking and viewing. This cost-effective framework strengthens ties with the school and community and provides students with the opportunity to understand real-world business issues and practices. Because most real-world problems do not fit neatly within the bounds of any one subject area, the curriculum tends to spread to interdisciplinary experiences. The English curriculum is integrated with various business enterprises as well as journalism and whatever technology is used in producing the promotional magazine.

A more detailed account of this project may be found in the May 1996 Journal of Adolescent and Adult Literacy, in an article entitled, "Students Utilize Their Multiple Intelligences to Develop Promotional Magazines for Local Businesses" by Jacqueline Glasgow and Margie Bush.

NCTE/IRA Standards

First, this activity requires students to read a wide range of print texts to build an understanding of texts, of themselves, and of the cultures of the United States and the world in order to acquire new information, to respond to the needs and demands of society and the workplace; and for personal fulfillment (NCTE/IRA Standard #1). In doing so, this multimedia project requires students to apply a wide range of strategies to comprehend, interpret, evaluate, and appreciate texts by drawing on their own experience, their interactions with other readers and writers, their knowledge of word meaning and of other texts, their word identification strategies, and their understanding of textual features (NCTE/ IRA Standard #3). In writing their articles and designing their pages, students will adjust their use of spoken, written, and visual language, e.g., conventions, style, and vocabulary to communicate effectively with a variety of audiences and for different purposes (NCTE/IRA Standard #4). To produce their magazine, students will use a variety of technology and information resources, such as digital cameras, scanners, and desktop publishing programs, to gather and synthesize information and to communicate knowledge (NCTE/ IRA Standard #8). And finally, students will use spoken, written, and visual language to accomplish their own purposes, e.g., for learning, enjoyment, persuasion, and the exchange of information (NCTE/IRA Standard #12) when they present their promotional magazine to the business owners and to the community.

Planned Performance Tasks

The classroom context for this multimedia presentation begins with a letter from a local business asking for the assistance of students in studying her small business. Her intent is to develop a promotional campaign for the business, including, specifically, a promotional magazine that she can place in prominent positions around town. See *Sample Inquiry-Based Letter to Launch the Exercise*.

Sample Inquiry-Based Letter to Launch the Exercise

TRACY APPLIANCES, INC.
2080 N. Cable Road
Lima, OH 45807
419–555–5550

19 August 1995

English 11 Advanced
Shawnee High School
3333 Zurmehly Rd
Lima, OH 45806

Dear Class,

My husband and I recently lost the advertising firm that has usually designed our brochures, written our advertisements for the local newspaper, TV and radio stations, and printed our brochures.

In light of this event and the creative work we know that you are doing in English class, we would like to give you the opportunity to engage in a real world exercise-one that is immediately applicable to the world of work.

Because we know that your age group is one of the most experienced as well as one of the most powerful consumer groups in the history of the retail industry, we want you to design a new marketing campaign for our store. To begin the process, we would like you to send one or two class representatives to the store to speak with us about what we will need. If you and your classmates agree to this project, we will introduce you to some of the small businesses around us, all of whom are in need of your assistance.

Please call me at the store by September 3, 1995.

Sincerely,

Marjorie Tracy

Marjorie Tracy
Tracy Appliances, Inc.

- Once the letters have been solicited and businesses have been identified, divide students in teams according to their interest in the various businesses they choose to study. Then, pass out copies of the project requirements. See *Promotional Magazine Requirements.*

Promotional Magazine Requirements

1. Front Cover
2. Title Page
3. Table of Contents
4. Two 500–word articles
5. Two 250–word articles
6. One article should be an interview with manager or owner
7. One should be a history of the business, its building, its product(s)
8. The other articles of either length are the team's choice: human interest, personal profile feature, topical events feature, news articles, letters to the editor, etc.
9. Two full-page, original advertisements for products peculiar to the business
10. Four photos or original drawings or computer clip art incorporated into articles, ads, or cover pages
11. Bibliography
12. Back Cover
13. Oral Presentation

Most teams exceeded the minimum requirements for the project and added extras. Some teams videotaped interviews with employees as well as the manager. Some teams opted to develop advertisements for radio and television. Most teams included more than two advertisements and four pictures.

- Prior to visiting the business they have selected, ask students to generate questions for interviewing the manager or key employee to create a business profile. Interviewing will be the primary source for gathering information for their projects. See *Student-Generated Interview Questions for Business Profile* for sample questions.

Student-Generated Interview Questions for Business Profile

1. What is your best selling product or service? Why did you choose these lines/services? What makes them good sellers?
2. How and when did this business get started? Whose idea was it? What need in the community was foreseen?
3. What is the future of this business? Do you plan to move? To change product lines?
4. Is this a recession-proof business? How do the fluctuations in the market and in the community affect your business?
5. How do you handle security? Do you screen your employees? What do you think about the practice of screening employees?
6. How do you handle such things as shoplifting, cheating?
7. How long has the business been in this location? Why did you locate here?
8. What is the history of the building?

- Also, prior to their first visit to the business, students need to generate interview questions for a personal profile of the manager or key employee of the business. See *Student-Generated Interview Questions for Personal Profile* for student examples.

Student-Generated Interview Questions for Personal Profile

1. How long have you been working at this job? For this business?
2. What kind of education and training does it take to do this job?
3. What changes have you seen occur in this job over the years?
4. What changes do you see taking place in the future for this business or in your occupation?
5. What do you like and dislike about this occupation? This business?
6. How do you balance a career and a family?
7. Is retraining or further education necessary for advancement in this job?
8. How does advancement occur in your profession?
9. What advice would you give to anyone who plans to enter this occupation?
10. Describe what you do during an average work day.
11. What kind of overtime or work after normal hours do you do?

- Students are now ready to arrange site visits to interview personnel, gather information, take photographs, and begin designing their promotional magazines. As students begin drafting their articles, provide instruction on writing the news story. Practice writing attention-getting headlines and strong leads that hook the reader into the story. Review the five W's and one H-who, what, when, where, why, and how-of a story.

Students may need instruction in writing the various types of feature articles: news, informative, historical, personality story, and first-hand account. Discuss an effective structure for an editorial. Refer students to *Writers INC: A Guide to Writing, Thinking, & Learning* (1992) for more information about writing these types of articles. See the *Student Feature Article* written by Julie Tracy about her parents' business.

Student Feature Article

Tracy's Appliances Remains in the Family
by Julie Tracy

Some forty years ago Harold and Marjorie Tracy started a Lima tradition, originally known as Tracy Refrigeration Service. Now under the name of Tracy Appliances the small family operated store has grown to be one of the areas leading appliance stores. Tracy Refrigeration Service started in a small building on the comer of Baxter and W. Wayne Street. This small white building had one window in the front with a door for shipping on the west side. At this time Marjorie Tracy handled all the book keeping, and phone calls on a single phone line. Soon 787 W. Wayne Street became too small for the rapidly growing business. In 1957, after six years at Wayne Street, the family built a 3,000 sq. ft. building in the 1700 block on Allentown Road. The seventies was a reconstruction phase for 1703 Allentown. In 1970 and 1972 the building was expanded to better service the growing needs. Jeff's handprint and name can still be seen on a cement block near the building. He put it there when he was only twelve after the final remodeling in 1972. The need for on site warehouse facilities, larger display rooms and bigger work spaces, soon left them no choice but to find a bigger place. In late 1991 it was determined that the old Tom Ahl car dealership, on Cable Road, would be a perfect new site for Tracy. Being next to the mall would provide the ability for consumers to ship in one convenient location. Tracy's moved into 2080 N. Cable Road in early 1992, and it is now their present day location. Family values have been an important part of the Tracy tradition. Tracy's is still owned and operated by the first generation of the Tracy family and hopes to continue the family tradition for many more generations.

- To prepare students to write an advertisement for their business, begin by asking them to bring in current advertisements similar to the business and products they are studying. Discuss headlines, layout considerations, illustrations, color enhancements, and appropriate company details. See the *Sample Student Advertisement* in the form of a jingle.

Sample Student Advertisement

Tracy's Version of Jingle Bells
adapted by Julie Tracy

Dashing through the crowds, people every place,
Up and down the aisles, sneezin' in my face!
It's so hard to choose, there's so much to see,
Wonder if what I got you, cost more than you got me? Oh!

Jingle bells, Jingle bells, will they never stop?
I've been shoppin' all week long and I'm about to drop. Oh!
Ring them bells, somewhere else, far away from here.
Ain't it really lucky that Christmas is only once a year? Oh!

Wrap your presents nice, pretty bows that shine,
Take them out to mail, you're gonna wait in line.
Fight your way back home, and if you are like me,
Maybe on the 24th, you'll get to trim the tree. Yet,

Jingle bells Jingle bells, good times without fail,
Oh, what fun it is to shop at Tracy's Truckload Sale.
Buy a fridge on a layaway get delivery that's free,
Open it on Christmas Day for all to share the spree!

Oral Presentations

Oral presentations were the culminating activity of this project. Each team prepared a 10 to 15 minute skit, demonstration, or speech that told the story of their research processes. Students prepared photographs, bulletin boards, videos, collages, music, and/or posters to enhance their presentations that we videotaped for assessment purposes and future viewing. (See *Assessment Rubric for Promotional Magazine Presentation* on page 104). In addition to presenting the biography of the project, students were also impressed by the commitment and time investment made to start and maintain a successful business. They learned the importance of dealing with people, meeting deadlines, and serving customers effectively. Money management, creative trends, and effective advertising were other important themes students discussed in their presentations. They found that running a business is not easy and that a good education is necessary to be successful. Most students came to a better understanding and appreciation of their parents, relatives and employers in the community where they live and work. One student expects to continue the family business started by his grandparents in 1928. See *Assessment Rubric for Promotional Magazine Project* on page 105.

Assessment Rubric for Promotional Magazine Presentation

	Exemplary (4)	*Proficient (3)*	*Developing (2)*	*Beginning (1)*	*Score*
Content	The content is superior in meeting the requirements of the assignment	The speaker provides enough content to meet the requirements of the assignment	The speaker does not provide enough content to meet the requirements of the assignment	The speaker says practically nothing	__ × 9 = (36 points)
Organization	The message is well-organized; the speaker provided transitions and clue words to indicate ideas, details, opinions	The message is organized and the sequence and relationships among ideas is clear	The organization of the message is mixed up; the organization appears to be random or rambling	So disorganized one cannot understand most of the message	__ × 6 = (24 points)
Speaking Skills	The speaker delivers the message in a lively, enthusiastic fashion	The pronunciation and enunciation are clear; appropriate volume and speaking rate	The audience must work to understand the message; the volume is too low and the speaking rate too fast	The pronunciation and enunciation are so unclear that one cannot understand most of the message	__ × 5 = (20 points)
Language	The speaker makes very few grammatical mistakes; the speaker uses language in effective ways to emphasize or enhance the meaning of the message	The speaker makes few grammatical mistakes; the speaker uses language that is appropriate for the task	The speaker makes many grammatical mistakes; the speaker uses very simplistic, bland language	The grammar and vocabulary are so poor that one cannot understand most of the message	__ × 5 = (20 points)

Total Points (100 possible)

Grade Scale
A =
B =
C =
D =

Assessment Rubric for Promotional Magazine Project

	Exemplary (4)	*Proficient (3)*	*Developing (2)*	*Beginning (1)*	*Score*
Content	Material is clear, relevant, accurate, and concise; unified, focused articles; details varied and vivid	Material is appropriate; topics or ideas generally clear; details generally varied and vivid	Material is usually appropriate; controlling topics and ideas lack focus	Little evidence of appropriate content	__×10=__ (40 points)
Organization/ Format	Organizes material in a clear, appropriate, and precise manner; creative expression	Organizes material in an appropriate manner, but lacks clarity or consistency	Little evidence of a cohesive plan; little or no description or detail	No evidence of a plan; ideas seem scrambled, jumbled, or disconnected	__×8=__ (32 points)
Creativity/ Inventiveness	Superior degree of originality throughout; high degree of elaboration; self-directed learning style	Average degree of originality throughout; shows ability to work and think independently	Below average degree of originality throughout; shows little initiative and independent thinking	Lack of originality; very little or no initiative; student waits to be told what to do	__×5=__ (20 points)
Research and Data Interpretation	Exceptional evidence of research; detailed facts; vivid descriptions	Substantial evidence of research; good amount of detail and description	Some evidence of research; some facts are accurate; some detail	No evidence of detail; incorrect or little facts; hardly any detail	__×5=__ (20 points)
News Articles	The ideas are clearly organized in the inverted pyramid format; logical progression; good attribution and nut graph; high quality graphics; captions provided for photos.	Ideas are there, but do not flow together or limited development of ideas; graphic images are relevant, but not unique. Need captions or captions not high quality	Strategies for organization are confusing. Insufficient examples; graphics are in the drafting stage	Not enough significant ideas from the novel; lacking direction and logical progression	__×10=__ (40 points)
Advertisements	Ads exhibit great skill in manipulation of media and techniques to persuade	Ads exhibit some skill in manipulation of media and techniques to persuade	Ads exhibit limited skill in manipulation of media and techniques to persuade	Ads exhibit little skill in manipulation of media and techniques to persuade	__×6=__ (24 points)
Graphics	Superior use of line, shape/ color, value, space, color, texture	Good use of line, shape/color, value, space, color, texture	Limited use of line, shape/color, value, space, color, texture	No regard for line, shape/color, value, space, color, texture	__×5 =__ (20 points)
Language	The language is vivid, imaginative, with correct and colorful word choice and description	Efficient use of language, but needs variety of sentence length and style	Insufficient variety of sentences; lacks sophistication of language usage	Word choices are inappropriate; there are many incorrect sentences	__×3=__ (12 points)
Conventions	Correct sentence structure, grammar, punctuation, and spelling	Mostly correct sentence structure, punctuation, grammar, spelling	Errors in sentence structure, punctuation, grammar, spelling	Many errors in sentence structure, punctuation, grammar, spelling	___×3=__ (12 points)

Total Points (187 possible)

Grade Scale
A =
B =
C =
D =

Bibliography

Glasgow, J., & Bush, M. (1996, May). Students use their multiple intelligences to develop promotional magazines for local businesses. *Journal of Adolescent and Adult Literacy,* 39, 638–649.

Sebranek, P., Meyer, V., & Kemper, D. (1992). *Writers INC: A guide to writing, thinking & learning.* Burlington, WI: Write Source Educational Publishing House.

10

THE TIME LINE OF THE TWENTIETH CENTURY

Linda J. Rice

An accomplishment sticks to a person.

Japanese Proverb

Overview of Authentic Learning

Integrating various disciplines and the language arts strands is easy to do with this unit, which helps students to develop their research skills, expand their critical thinking capacity, and utilize a variety of resources and their own creativity to collaboratively create a Video Masterpiece. The unit encourages individual accountability as students become "expert contributors" on a particular area of their group's assigned era of the Twentieth century, and cooperative learning where students pool their knowledge in order to create a video that, in effect, presents or teaches highlights of their era to the rest of the class. In order to make this work, students are assigned to (or choose) groups to cover the following eras:

- Turn of the Century and World War I (1901–1918)
- The Roaring Twenties to the Great Depression (1920–1939)
- World War II (1939–1945)
- The Civil Rights Movement and Feminism (1950–1969)
- Vietnam (America's Involvement, 1962–1975)
- The Eighties and Nineties (1980–1999)

Once the groups are formed, each student in the group will choose one aspect of their assigned (or chosen) era to research. This is the area in which the student is to become the group's "expert." The areas are as follows:

- Political and Economic Personalities and Events
- Military Events
- Scientific and Technological Discoveries and Advancements
- Sociologic Events (social needs/problems, interpersonal relationships, statistics about the daily lives of people) and Entertainment

NCTE/IRA Standards

In this unit students will:

- Conduct research, gather, synthesize, and communicate information (NCTE/IRA Standard #7) highlighting a variety of issues, discoveries, and people influential in the twentieth century. Students will use a variety of technological and informational resources to complete this task (NCTE/IRA Standard #8).
- Read a wide variety of print and nonprint texts to build an understanding of the cultures of the United States and the world (NCTE/IRA Standard #1).
- Apply a wide range of strategies (such as reading, note taking, editorial writing, artifact finding, poster making, speech writing, group deliberating, and video taping/editing) to comprehend, interpret, evaluate, and appreciate the texts of life, history, and entertainment in the twentieth century (NCTE/IRA Standard #12). In this process, students will draw on their knowledge of other texts and interactions with other researchers (NCTE/IRA Standard #3).
- Employ a wide range of strategies, from note-taking and editorial writing to speech writing and outlining, to communicate with different audiences for a variety of purposes (NCTE/IRA Standard #5). In doing so, students will apply their knowledge of language structure and conventions (NCTE/IRA Standard #6).
- Participate as knowledgeable, reflective, creative, and critical members of a research community whose task is to communicate accurate and interesting information (NCTE/IRA Standards #4 & #11) about an era of the twentieth century to the rest of the class. Because students will accomplish this task through a video, students must also apply their knowledge of media techniques (NCTE/IRA Standard #6).

Performance Tasks

Performance Task for Note Cards and Documentation of Research

Although the culmination of this unit is the Video Masterpiece, there are numerous steps that lead to it, the note cards being foundational. Thus, the unit begins as students, each having a pack of 100 3 x 5 note cards, journey to the library or media center to gather facts about their assigned area of their group's assigned era. The goal is for each student to find at least 25 important pieces of information about her/his area to contribute to the group's Video Masterpiece, which is to provide a comprehensive look at their particular era of the twentieth century.

In addition to needing 25 pieces of information, students should fill all 100 cards. Except for cards used to cite the MLA documentation for the source of the information, all other cards should contain detailed written notes. Some pieces of information may only take one note card, whereas others may have supporting details and information requiring more than one card. For instance, students investigating the Clinton Impeachment would very likely have 4 to 10 cards outlining various aspects of the hearings, its key players, and outcomes.

In order to maximize students' research capabilities (because there are many books that could be used as the sole source of information, and to allay students' temptation to look up all of their information online, at home, and therefore make poor use of class time), they should be required to utilize at least seven different sources representing at least four different types. Books, periodicals, CD-ROMS, and online sources all ought to be required. And, if a public library with microfiche/microfilm is available, students (college-bound, in particular) would benefit from utilizing those resources as well, especially to see the way the news was reported at the time the influential events actually happened.

Performance Task for the Editorial

Having identified 25 pieces of information on their individually assigned areas of their group's assigned era, students now have the task of deciding which of these pieces of information are the most important. Therefore, from the possible list of 25 topics, students will choose the one they find to be the most influential of their area of their era. The "top choice" could be an event, a person, an idea, invention, issue, or happening.

After making the "top choice," the student will write a four to six page editorial that includes the following:

- Background information on the topic
- Statement of the significance then
- Statement of its lasting influence/effect

- Explanation of how learning about this topic influenced the student's understanding of their area and era

The editorial should be prepared according to MLA style and written in a well-organized fashion that logically integrates research from at least 4 sources. See *Assessment Rubric for Editorial*.

Assessment Rubric for Editorial

	Exemplary	*Acceptable*	*Needs Improvement*
Coverage, Development, and Significance	Addresses topic in a logically organized, well-transitioned manner; clearly establishes both the impact then and now, including how the event has affected the student's own understanding of the area and era; 4 to 6 pages	Addresses topic but either lacking in overall organization and transitions to link ideas, or showing only partial impact, so as to leave out the "then" or "now" or how knowledge of the "top 3" have affected the student's own understanding; 4 pages	Wanders without focus on one influential event, person, idea, invention, issue, or happening; 3 or fewer pages
Resources	4 sources representing at least two types; well-integrated into the paper	3 sources representing at least two types; well integrated into the paper	3 sources (or fewer), but all from Internet
MLA Style	Complete adherence	1 to 2 errors	3 or more errors
Grammar Mechanics, and Usage	One or no errors per page	2 to 3 errors per page	More than 3 errors per page

Performance Task for Editorial and Artifact Sharing

Now that students have surveyed (as demonstrated by their 100 note cards) their assigned area of their group's assigned era and composed editorials indicating their "top choices" of influential events, persons, ideas, inventions, issues, or happenings, they are to find, or create a model of, an artifact representing their area and era, preferably related to the editorial as well. For a list of artifacts students have shared, see *Artifacts for Sharing* on page 112.

On the day scheduled for editorial and artifact sharing, the class sits in roundtable fashion. Ideally, group members sit next to each other and in order. This way, the sharing becomes a kind of living artifact time line where students can see important developments spanning the twentieth century. For time/practicality sake, students may want to either verbally summarize their editorials or read "essential highlights" (rather than reading all of the four- to six-page paper). It is also nice to pass artifacts around the circle after the presenter tells of its significance, though particular care should be given to items that are delicate from age and/or have high sentimental or material value. (See *Assessment Rubric for Artifact Sharing* on page 112.)

Performance Task for Speech on Autobiography or Biography of Choice

To foster a more in-depth look at some of the key "movers and shakers" of the twentieth century (see *Possibilities for Autobiography/Biography, "Movers and Shakers" of the Twentieth Century* on pages 113 and 114), this part of the Time Line unit requires each student to choose and read an autobiography or biography of an influential person from their group's era. For practical purposes, the autobiography or biography does not have to also represent the students' assigned areas of research. Therefore, students whose assigned area is Science and Technology will have the same freedom to read about a famous entertainer that the person assigned Sociology and Entertainment would have.

Once the students have had a couple of weeks to read their autobiographies or biographies, they will prepare a speech accompanied by a poster that gives visual cues about the influential person. Students may use note cards to assist them with their delivery; however, their primary eye contact should be with their peer audience. The speech should be four to six minutes long and will be assessed (see *Assessment Rubric for Speech on Autobiography or Biography of Influential Person* from Assigned Era on page 115) half in terms of delivery and half in terms of content. The content should be organized in a logical manner and filled with enough interesting and/or important anecdotes about the person's contributions and/or way of living to keep the attention of the listeners. The poster should also be deliberately incorporated into the speech by alluding to its content at relevant times during the speech, rather than tacking it on at the end.

Artifacts for Sharing

The artifacts listed below indicate what one class brought in during the artifact sharing portion of the unit presentations.

Turn of the Century and World War I (1901–1918)

- Egg beater
- Grandmother's baby doll
- Barber tools (hair clipper, shavers, shaver sharpeners, slate and belt)
- Photograph of Great Aunt feeding chickens
- Photograph of Great Grandpa in army uniform with weapons
- Gunter's chain used for surveying

World World II (1939–1945)

- Navigators Flight Book/Information File
- Photograph of Great Step Uncle in navy uniform
- Ration book
- Poster (reproduction) of "Rosie the Riveter"
- Life magazine featuring Winston Churchill

The Civil Rights Movement and Feminism (1950–1959)

- First Beatles album
- Box of 45s (45 RPM records)
- Advertisement for Volkswagen Beetle
- Post cards from Disneyland
- Sports page with article about Satchel Paige

Vietnam (America's Involvement, 1962–1975)

- Hippie belt
- Nixon political pin
- Lava lamp
- Commemorative stamps issued after JFK's assassination
- Tie dye T-shirt

The Eighties and Nineties (1980–1999)

- Princess Diana doll and book, along with song Elton John sang at her funeral
- Ammunition from Desert Storm
- CD player
- "Just Say No" button
- Ross Perot political pin
- MTV video

Assessment Rubric for Artifact Sharing

	Acceptable	*Needs Improvement*
Vocal Production	Sufficient volume; clear, articulate voice that conveys interest (sincerity and/or enthusiasm); fluid, demonstrating that the reading is rehearsed or that the "talk" has involved advance thought and preparation	Student could benefit from increased focus on: ___________* here teacher would fill in some area from the "acceptable" column that needs more attention
Relevance of Artifact	Clearly established and linked with their area of their era	From the era but not the assigned area

Performance Task for Video Masterpiece

Now that students have become individual "experts" on their area of an era, they are ready to work with their group members to create the Video Masterpiece. In its entirety, each group's video should be 20 to 25 minutes in duration. Required components of the video are as follows:

- Ten pieces of information from each area/student
- At least 3 musical selections from the era
- At least 3 film clips from the era (The Roaring Twenties and Great Depression must substitute one of the film clips for a radio show from the era)
- An interview with a person from (or knowledgeable about) the era
- At least one scene acted out for each area (e.g., the moon landing, JFK's assassination, the discovery of the polio vaccine)

Students will want to consider a variety of ways to structure the video and integrate the required elements in a meaningful, cohesive fashion. For instance, students may use the anchor man/news show format or an Oprah-style interview of experts on the era. To facilitate students' task of organizing, each group will prepare an outline of the information covered, complete with listings of techniques used (music selection, film clip, interview, acting, artifact sharing, photo, commercial break, sound bite from famous people, etc.). Instructions for the outline are explained by *Instructions for Outline to Accompany Video Masterpiece* on page 116. See also *Assessment Rubric for Video Masterpiece* on page 117.

Possibilities for Autobiography/Biography, "Movers and Shakers" of the Twentieth Century

This list was prepared by Lakeview High School's Media Specialist, Chris Daubenspeck, to indicate what books currently exist in our library for this portion of the Time Line of the Twentieth Century unit. Students may also go to the book store or public library to select an autobiography/biography; however, it is a good idea for the teacher to "pre-approve" the students' selections to ensure that the books are at the desired level of challenge. Otherwise, some students may be tempted to download their information from the Internet, watch a video, or simply read a children's book. It is recommended to actually set a due date and even give a "head start reading day" to ensure that the students have actually selected a book on which to base their speech.

Media/ Entertainment/Sports

1920s

- Billie Holiday

1930s

- Josephine Baker
- Lou Gehrig
- Martha Graham
- Vivien Leigh
- Babe Ruth

1940s

- Jack Benny
- Irving Berlin
- Jackie Robinson

1950s

- Miriam Anderson
- Roy Campanella
- Helen Hayes
- Satchel Paige

1960s

- Judy Collins
- Joe DiMaggio
- Walt Disney
- Aretha Franklin
- Vince Lombardi
- Leonard Nimoy
- Gary Player

1970s

- Robert Clemente
- Sugar Ray Leonard

1980s

- Christopher Reeve

1990s

- Mark McGwire

Science/Exploration

1920s

- Sigmund Freud
- Charles Lindbergh

1930s

- Margaret Mead

1940s

- Charles Yeager

1950s

- Linus Pauling
- Werner Von Braun

1960s

- Robert Hutchings Goddard

1980s

- Christa McAuliffe

1990s

- Jocelyn Elders

Business

1920s

- Howard Hughes

1930s

- Henry Ford

1970s

- Lee Iacoca

1980s

- Steve Wozniak

Politics/Military

1930s

- Franklin and Eleanor Roosevelt

1940s

- Douglas MacArthur
- General George Patton

1950s

- Dwight D. Eisenhower
- Herbert Hoover
- Sandra Day O'Conner
- Rosa Parks
- Adlai Stevenson

1960s

- Spiro Agnew
- Fidel Castro
- Jane Fonda
- Lyndon B. Johnson
- John F. Kennedy
- Robert F. Kennedy
- Martin Luther King, Jr.
- Nikita Khrushchev
- Richard Nixon
- Malcolm X

1970s

- Jimmy Carter
- Jesse Jackson
- Gloria Steinem

1980s

- Ronald Reagan
- Ryan White

1990s

- George Bush
- Bill Clinton
- Sadam Hussein

Literature/Art

1920s

- George Bernard Shaw

1940s

- Carl Sandburg

1950s

- Lorraine Hansberry
- Ernest Hemingway
- John Steinbeck

Assessment Rubric for Speech on Autobiography or Biography of Influential Person from Assigned Era

	Exemplary	*Acceptable*	*Needs Improvement*
Content (50% of speech grade)	Strong sense of organization and logical development as the speech unfolds to show a depth of understanding about the person's life; Rich with interesting information and anecdotes that really engage the class	Shows depth of understanding, but is either presented in a wandering fashion or includes only a few (2 to 3) anecdotes; focuses on the "basic" facts of the person's life, reading more like a resume than a real, vital, interesting, three-dimensional person	Only a superficial coverage of the person's life and accomplishments or containing only 1 anecdotal/narrative type story that keeps the attention of the audience
Delivery Components (50% of speech grade)			
Vocal Production	Volume is sufficient throughout; voice is varied in characteristics to build intensity and draw the audience through the relevant emotional range of the person's life; clear articulation	Volume is sufficient overall, but flags 1 to 2 times during the speech; voice is sincere, but not particularly varied to convey interest, range, or intensity consistent with the person's life; some moments of unclear speech/articulation	Difficult to hear-serious need to increase volume; flat, monotone vocal qualities that challenge the audience's ability to pay attention with ease
Eye Contact and Facial Expression	Sees everyone in the class; looking up 80% of the time, consulting notes only periodically; no sense of "reading" to the class; lively and/or sensitive in ways that show understanding of content and therefore facilitate meaning	Tendency to direct most eye contact either to the teacher or a small number of students; Heavy reliance on notes, but still not "reading" to the class; limited expressive range in terms of eyes and facials helping to convey meaning	Mostly reading the speech to the class; rarely looking up to see the audience
Gesture and Body	Poised throughout the speech; walking on transitional sentences; extends gestures to draw in the audience; keen awareness of gestures that facilitate, not distract, overall meaning and impact of words and ideas shared	Poised throughout the speech, but planted in one place; some gestures are included, but tend to be held close to the body, rather than extended to draw in audience and maximize facilitation of meaning	Noticeably shifting weight back and forth and/or hands in pockets; no deliberate gestures
Sense of Spontaneity and Newness	Consistently strong, sincere energy level; reflects a genuine interest in what is being shared	Strong overall, but periodic lapses that detract from the overall energy of the speech	Monotone or bored sounding
Flow of Speech	Tempo works in conjunction with meaning throughout the entire speech; varied as necessary to draw emphasis; effective use of pause	Tempo is varied at times, indicating some awareness of its effect on the build of the speech, but to a lesser degree than the "exemplary" description	Rushed or laborious; Not varied

Instructions for Outline to Accompany Video Masterpiece

All Groups: On the day your video is due, you must turn in an outline organized in the order of your video. The outline must include the following:

- 10 pieces of information per person in your group
  - each piece of information should be followed by the name of the person who contributed it and be numbered 1 to 10
  - also, where applicable, the technique used to present the piece of information should follow the contributing student's name)

    e.g., Snow White and the Seven Dwarfs debuted in 1937 (Natalie 1, film clip)
- 3 video clips (list the movie title and what year it came out)

  e.g., Gone with the Wind, 1932
- 3 songs (song title and performing artist/group must be included)

  e.g., "We all live in a yellow submarine," The Beatles
- scenes (one per area/person in your group) you acted out

  e.g., JFK's assassination, December 1963.
- interview(s) with person's name, relation, and who conducted the interview

  e.g., Mr. Kirnshaw, Katie's uncle, a Vietnam veteran (interviewed by Katie and Michael)

Format for Outline

- MLA Header
- Double-spaced

Numbered 1 to 50 (or however many numbers it takes to cover the information above)

Bibliography

Gibaldi, J. (1999). *MLA handbook for writers of research papers* (5th ed.). New York: The Modern Language Association of America.

Assessment Rubric for Video Masterpiece

	Exemplary	*Acceptable*	*Needs Improvement*
Coverage of each area	10 pieces of information per person in group	7 to 9 pieces of information per person in group	6 or fewer pieces of information per person in group
Video clip	3 films of or representative of the era	2 films of, or representative of, the era	1 clip of, or representative of, the era
Songs	3 songs from the era	2 songs from the era	1 song from the era
Scenes acted by group	One from each area of research; acted in a manner (be it fun or somber) consistent and respectful of the content	Missing one area, but otherwise meets description listed under "exemplary"	Missing 2 areas or acted in a way that is inconsistent with the seriousness of the event (i.e., joking around about an assassination)
Interview	Conveys meaningful information about the era from an "informed" person; close-up filming and easy-to-hear	Information is somewhat tangential to the era, thus causing audience to wonder about the quality of the questions that were asked; filmed from a distance or difficult to hear	None
Organization and integration of information	Well-organized and merging various required components in a logical way that maximizes meaning and impact	Strong sense of organization and integration overall, but periodic randomness or "listing" effect, such as playing all three films in a row, rather than interspersing them with commentary and/or other information and presentation techniques such as music and acting	Appears as though each individual did his/her own part of the video without any real sense of collaboration or effort to integrate the various components in a logically compelling and/or visually interesting way
Motif, fluidity and quality of video production	Binding motif (such as Oprah-style show, "TV Decades Special," or anchor desk concept) and careful editing combine to make the video flow in a professional fashion with close-ups and clear volume	Similar to the "exemplary" description, but with 2 to 4 brief technical problems (such as editing gaps, unclear sound and/or picture quality) that distract from the content	More than 4 technical problems or lack of binding motif to unify the presentation as a whole

Part III

Oral Performance Assessments

11

POLICING SPEECH: BRINGING AN OPEN FORUM TO THE CLASSROOM

Ruth McClain

"The man who does not read good books has no advantage over the man who cannot read them."

Attributed to Mark Twain

(Adapted from an idea by June Berkley and Christie Simmons, Ohio University)

Overview of Authentic Learning

Not only are schoolbook censors active everywhere, they are well-funded and organized. Challengers of freedom to learn often describe their activities as a legitimate exercise of parental rights, and, certainly, they are free to request that their own children be excused from specific activities. But, when they bring such arguments in support of efforts to remove certain literature from a curriculum so that no child may use that literature, this is censorship.

Attackers of freedom to learn know few bounds, and, in attacking freedom of expression, challengers teach that ideas are things to be feared, that intolerance is a desired trait, and that diversity is a source of trouble.

School censorship, particularly, is persistent and unpredictable. It knows no geographic or demographic boundaries. For teachers, then, the best preparation for challenges is vigilance and the practice of democracy in the classroom, where all ideas can be freely shared and appreciated. In an effort to establish such a democratic classroom, teachers must engage students in "real" literature and offer them a forum for freedom of expression. This assessment requires students to write a rationale for a controversial Young Adult novel and then participate in the Open Forum to defend their novels.

NCTE/IRA Standards

In this assessment, students will:

- read and respond orally and in writing to a wide variety of texts representing a variety of authors, cultures, and eras (NCTE/IRA Standard #1).
- apply a wide range of strategies to comprehend, interpret, and evaluate texts (NCTE/IRA Standard #3).
- research a controversial Young Adult novel. In doing so, students will conduct research on issues and interests by generating ideas and questions, and by posing problems. They will gather, evaluate, and synthesize data from a variety of sources to communicate their discoveries in ways that suit their purpose and audience (NCTE/IRA Standard #7); and students will use a variety of technological and informational resources to gather and synthesize information and to create and communicate knowledge (NCTE/IRA Standard #8).
- communicate effectively for a variety of audiences, situations, and purposes (NCTE/IRA Standard #4); students participate as knowledgeable, reflective, creative, and critical members of a variety of literacy communities (NCTE/IRA Standard #11); and, students will use spoken, written, and visual language to accomplish their own purposes (NCTE/IRA Standard #12).

Performance Tasks for Developing a Rationale for a Controversial Young Adult (YA) Novel

This project has several purposes and components related to teaching Young Adult novels, selecting texts, and writing rationales to help prevent questions being raised about the literature you select for your students to read.

- Select a Young Adult novel from the booklist.
- Write a Rationale using Second Language Acquisition and Teacher Education (SLATE) Guidelines for Writing a Rationale. See *Books on 1999–2000 Censorship Lists*. Also see on page 124, *Excerpts from a Rationale on the Potential Problems with the Harry Potter Series* by Amy Weyand and on page 125, *Excerpts from a Rationale on the Redeeming Values of the Harry Potter Series.*

Guidelines for Writing a Rationale

(Adapted from SLATE Starter Sheet, NCTE, April 1994)

- The bibliographic citation
- The intended audience
- A summary of the plot
- Potential problems with the work and ways to solve them
- Redeeming values: Why should the book not be banned?
- List and summary of book reviews
- Objectives, teaching methods and assignments
- Alternative works an individual student might read and why
- Sources that recommend the book
- Awards earned by the author

Books on 1999–2000 Censorship Lists

- *Harry Potter Series* by J. K. Rowling
- *We All Fall Down* by Robert Cormier
- *In the Middle of the Night* by Robert Cormier
- *Fade* by Robert Cormier
- *Fallen Angels* by Walter Dean Myers
- *Go Ask Alice* by Anonymous
- *Annie on My Mind* by Nancy Garden
- *The Boy Who Lost His Face* by Louis Sachar
- *Snow Falling on Cedars* by David Guterson
- *Chinese Handcuffs* by Chris Crutcher
- *Athletic Shorts* by Chris Crutcher
- *The Crazy Horse Electric Game* by Chris Crutcher
- *Bud Not Buddy* by Christopher Curtis
- *Sarny* by Gary Paulsen
- *When She Hollers* by Cynthia Voigt
- *Tangerine* by Edward Bloor
- *Whirligig* by Paul Fleischman
- *Crusader* by Edward Bloor
- *Welcome to the Ark* by Stephanie Tolan
- *Gathering Blue* by Lois Lowry
- *Deliver Us from Evie* by M.E. Kerr
- *Hello, I Lied* by M.E. Kerr
- *Wringer* by Jerry Spinelli
- *All We Know of Heaven* by Sue Ellen Bridgers
- *The Summer of My German Soldier* by Bette Greene

Excerpts from a Rationale on the Potential Problems with the *Harry Potter Series*

Potential Problems with Harry Potter

In a country that prides itself on free speech, censorship has become a hot topic, a topic that has been portrayed in negative light in such books as Fahrenheit 451 by Ray Bradbury. Harry Potter has seen his share of negative publicity and according to the American Library Association Rowling's books were the most challenged in America. So why all the fuss? As discussed below, parents are concerned at the lack of parental control, themes of evil, and use of spells and witchcraft. For any teacher to use the books in their classroom they must be prepared to defend their decision. If I were ever to use the book I would first send a letter home to parents warning them that I will be using the book. In the note I would also encourage all parents to read the book before forbidding their children from reading the books. I believe that if people give the books a chance they too will fall in love with Harry Potter.

- Witchcraft: Harry and his friends are learning how to become witches and wizards. During this education they are taught that witchcraft is something that most people cannot do and something that needs to be taught and watched by adults. Also, the witchcraft is not evil if administered by most of the people in the book. If someone was evil they were viewed as evil, not good.
- Evil: There are elements of evil in the book. Yet, the evil is never glorified or shown in a good manner. Anyone who is seen as evil is not someone that any reader would like, anyone that a reader would aspire to be. Rowling is showing that evil is not a good thing at all.
- Spells: Throughout the book spells are used. But, spells are never allowed without supervision of an adult. So, if a child would want to be like Harry they would understand that magic is something that must be understood at a high level before using it on their own.
- No Adult Control: When parents notice little adult control in a book they become worried. Although there is little control Harry and his friends always act appropriately. They are kind to one another and are role models for great friendships. Rowling is not glorifying children without adults.
- It's Scary: Harry Potter and his friends encounter some fairly frightening characters. But, they are not so scary that children will be frightened. In all fairy tales there is a scary, bad person but children are also able to tell fact from fiction.

Excerpts from a Rationale on the Redeeming Values of the Harry Potter Series

Harry's Redeeming Values—Why the Book Should not Be Banned

In a review that ran in the New York Review of Books, it was said that children identify with a character like Harry who feels left out from the world around him. When Harry is with muggles he never feels right, but once he is in the world of magic he finds a place for himself in a world of his own. This is a strong feature of Harry Potter: helping children realize that at some point they will be able to find a place for themselves in the world, a place where people will accept them for who they are.

Since Harry Potter and the Sorcerer's Stone took America by storm children have started to love to read. Children that never would have sat down to read a short book are reading Rowling at lightning speed, even holding reading contests with their friends to see who can read the fastest. And not only are children merely reading but they are reading something worthwhile, something that will teach them valuable life lessons. In an article that ran in the ALAN Review it was said that the book's main characters stand up for their friends and for people that are not as fortunate as them. Aren't these redeeming values? These values are lessons that any parent would want their children to learn. Also in this ALAN Review article the author praises the "sensitivity and caring" that Harry and his friends show towards one another and the people around them. In short, Rowling is not only teaching values to be used as children, but also life values that will help her readers become better people.

Harry's many journeys are classical tales with new twists that allow children to learn more about classic storytelling. Harry's fights with evil never come easy and Rowling does not play down the danger that he faces. Children are taught that one must not give in to "the dark side." Rowling's Potter is merely Star Wars for children of this generation, a wonderful story that will draw young readers in to reading and to become lifelong readers.

Throughout Harry's many adventures Harry is held responsible for his actions. This emphasis on responsibility also runs through the practice of witchcraft throughout the book. Rowling is not advocating that children run out and practice witchcraft but that only properly trained people can do the things Harry does. And the fact that Harry excels in school is also a redeeming value of the book. Rowling's characters are not perfect although they each excel at something special. This is an important lesson for many students. While Harry does not give up on education he also is a whole person, a valuable person. This is a great aspect for kids who might not excel at school, that they will still find something that fits them in life.

Creativity and imagination are valued assets in American society. People that excel the most are those who think of new ways to do normal, everyday things. Harry Potter books force readers to imagine worlds they will never see and things they will never experience. These qualities are another positive aspect of the book: forcing children to imagine.

Assessment Rubric for Developing a Rationale

Components	*Quality*	*Skilled*	*Acceptable*
	5 points	**4 points**	**3 points**
1. *Plot Summary*	The summary has a clear focus with elaboration of supporting facts and details.	The summary has an ambiguous thesis and is lacking supporting details.	The summary contains the main events of the novel, but the focus is unclear and supporting details are lacking.
2. *Potential Problems and Ways to Address Them*	Clearly articulates potential problems with this book and gives concrete ways to address them.	Problems are identified, as well as generic reasons for teaching this book. Needs more significance or reliability to convince me.	Ambiguous identification of problems; the redeeming values given for this book are not persuasive, but would be acceptable if the book is not challenged.
3. *Redeeming Values; Why the Book Should not be Banned*	Articulates the redeeming values of the book and significant reasons why this book should be taught; makes connections to students' lives. Uses citations to give authority to your position.	Identifies the redeeming values; doesn't necessarily connect to students' personal lives; makes some connections to criticism.	Makes reference to the values of the book, but doesn't connect with students' lives and citations are missing.
4. *List and Summary of Reviews*	Reviews come from reliable sources; they highlight the controversial aspects of the book; they are correctly cited with brief discussion.	Reviews come from credible sources, but discussion is limited or critical aspects of the controversy are omitted.	Reviews come from questionable sources; they lack critical aspects of the controversy; they may have errors in the citations.
5. *Objectives, Teaching Methods and Assignments; Alternative Reading Assignments*	List of suggested teaching objectives; list of suggested students activities; possible essay questions; list of vocabulary words; alternative reading/related book lists are provided with justification for the choices.	Missing components or description not fully developed; alternative reading/related book lists are provided, but reasons for substitution are not clear or list not inclusive.	Lacking essential components; insufficient information; alternative readings are missing or reasons lack conviction.
	3 points	**2 points**	**1 point**
6. *Sources That Recommend the Book*	Sources are reliable and credible with clear recommendations quoted.	Sources are reliable, but quotes lack substance.	Sources are limited and recommendations are ambiguous or generic.
	2 points	**1 point**	**0**
7. *Awards Earned by the Author*	Persuasive list of awards to justify this author's work. Bibliography of works cited.	Convincing list of awards to justify this author's work.	Incomplete or missing a list of awards by this author.

Performance Tasks for the Open Forum

Divide the class into three different groups:

1. Group 1 is pro-censorship
2. Group 2 is anti-censorship
3. Group 3 is a group of concerned citizens from the community —some of whom are neutral, some angry, some liberal, etc.

- ♦ Groups 1 and 2 are given the following instructions:

  Students are to assume the roles of the following persons (ask students to interview these people and find out the various positions they take on censorship):

  - school superintendent
  - school principal
  - school librarian
  - classroom teacher
  - parent of a student in whose class the controversial piece was taught
  - student in the class

  (Each group of students is then sent out to interview the various persons whose role they have assumed. They formulate specific questions stating that the controversy centers around a specific piece of literature taught in a specified grade level or class.)

  (Each group does this and relates to the interviewee whether he/she must assume a positive or negative approach to the piece in question. Students then meet in their own groups to formulate their strategies and anticipate questions that may be asked during the forum.)

- ♦ Group 3 is subdivided into two smaller groups—one of which is an angry group of parents who have come to the forum to protest the use of the literature in question. The other is a group of parents who are far more liberal and completely support the First Amendment. These two groups meet separately and formulate a set of objections or supports that they will then bring to the forum.

The teacher role plays moderator. The forum can be videotaped. Guests from the community and school can be invited to sit in the audience from which questions will be asked. The local newspaper should also be invited and asked to editorialize the forum. See *Assessment Rubric for the Open Forum* on page 128.

Assessment Rubric for the Open Forum

Exemplary (4)	*Adequate* (3)	*Minimal* (2)	*Unacceptable* (1)	*Score*
Weighs multiple perspectives on an issue and considers the good; uses relevant knowledge to analyze an issue; employs a higher order discussion strategy such as argument by analogy, stipulation, or resolution	Demonstrates knowledge of important ideas related to an issue; states an issue for the group to consider and presents more than one viewpoint; supports a position with reasons or evidence	Makes statements about the issue that express only personal attitudes; mentions a potentially important idea but does not pursue it in a way that advances the group's understanding	Remains silent or contributes no thoughts of his or her own; makes only irrelevant comments	__ × 15 = (60 points)
		Procedural		
Engages in more than one sustained interchange or summarizes and assesses the progress of the discussion; makes no comments that inhibit others' contributions and intervenes if others do this	Engages in an extended interchange with at least one other person; paraphrases important statements as a transition or summary; asks another person for an explanation or clarification germane to the discussion; does not inhibit others' contributions	Invites contributions implicitly or explicitly; responds constructively to ideas expressed by at least one other person; tends not to make negative statements	Makes no comments that facilitate dialogue; makes comments that are primarily negative in character	__ × 10 = (40 points)

Total Points:

Grading Scale:

A =

B =

C =

D =

F =

Bibliography

Anonymous (1994). *Go ask alice.* New York: Simon & Schuster.

Bloor, E. (1999). *Crusader.* New York: Harcourt Brace.

Bloor, E. (1997). *Tangerine.* New York: Harcourt Brace.

Bridgers, S. E. (1996). *All we know of heaven.* Wilmington, N.C.: Banks Channel Books.

Cormier, R. (1991). *We all fall down.* New York: Delacorte.

Cormier, R. (1999). *In the middle of the night*. New York: Econo-Clad Books.

Cormier, R. (1988). *Fade.* New York: Delacorte.

Crutcher, C. (1996). *Chinese handcuffs*. New York: Dell Publishing.

Crutcher, C. (1991). *Athletic shorts.* New York: Greenwillow Books.

Crutcher, C. (1987). *The crazy horse electric game.* New York: William & Morrow.

Curtis, C. (1999). *Bud, not buddy.* New York: Delacorte Press.

Fleischman, P. (1998). *Whirligig.* New York: Henry Holt.

Garden, N. (1982). *Annie on my mind.* New York: Farrar, Straus, Giroux.

Greene, B. (1974). *Summer of my German soldier.* New York: Bantam Press.

Guterson, D. (1994). *Snow falling on cedars.* San Diego: Harcourt Brace.

Kerr, M. E. (1986). *Deliver us from Evie.* New York: HarperCollins.

Kerr, M. E. (1997). *Hello, I lied.* New York: HarperCollins.

Lowry, L. (1993). *Gathering blue.* New York: Houghton Mifflin.

Myers, W. D. (1988). *Fallen angels.* New York: Scholastic Books.

Paulsen, G. (1997). *Sarny.* New York: Delacorte Press.

Rowling, J. K. (1999). *Harry Potter and the prisoner of Azkaban.* New York: Scholastic.

Rowling, J. K. (1999). *Harry Potter and the chamber of secrets.* New York: Scholastic.

Rowling, J. K. (1998). *Harry Potter and the sorcerer's stone.* New York: Scholastic.

Rowling, J. K. (2000). *Harry Potter and the goblet of fire.* New York: Scholastic.

Sachar, H. (1999). *The boy who lost his face.* New York: Econo-Clad Books.

Spinelli, J. (1997). *Wringer.* New York: HarperCollins.

Tolan, S. (1996). *Welcome to the ark.* New York: Morrow Junior Books.

Voigt, C. (1994). *When she hollers.* New York: Scholastic.

12

POETRY COFFEE HOUSE

Janie Reinart

We must learn to see the world anew.

Albert Einstein

Overview of Authentic Learning

If we are serious about building tolerance, care, and social justice, we must incorporate more of these materials into the curriculum. Available to teachers and young adults is a wealth of poetry anthologies written and/or edited by modern authors such as Jo Carson, Lori Carlson, Gary DeSoto, Mel Glen, Paul Fleischman, and Paul Janeczko. Students can choose a social issue they are interested in pursuing and create a poetry anthology of their own that includes poems they have collected, analyzed, and illustrated, as well as original poems they have written. Students must also research their social issue finding current information in newspapers, magazines, and Internet resources. The culminating activity for this assessment requires students to perform their poems at a Poetry Coffee House where they receive feedback from an invited guest or local poet.

NCTE/IRA Standards

In this assessment, students will:

- read, collect and analyze a wide range of poetry in order to create their own poetry anthology. In doing so, students will read a wide range of texts to build an understanding of texts, of themselves, and of the cultures of the U.S. and the world; to acquire new information;

to respond to the needs and demands of society and the workplace; and for personal fulfillment. Among these texts are fiction and nonfiction, classic and contemporary works (NCTE/IRA Standard #1); students will read a wide range of literature from many periods in many genres to build an understanding of the many dimensions (e.g., philosophical, ethical, aesthetic) of human experience (NCTE/IRA Standard #2); and students will apply a wide range of strategies to comprehend, interpret, evaluate, and appreciate texts. They draw on their prior experience, their interactions with other readers and writers, their knowledge of word meaning and of other texts, their word identification strategies, and their understanding of textual features (NCTE/IRA Standard #3).

- write and perform various types of poetry for their poetry anthologies and for the Coffee House performance. In doing so, students will use a variety of technological and informational resources (e.g., libraries, databases, computer networks, video) to gather and synthesize information and to create and communicate knowledge (NCTE/IRA Standard #8); students adjust their use of spoken, written, and visual language to communicate effectively with a variety of audiences and for different purposes (NCTE/IRA Standard #4); students will employ a wide range of strategies as they write and use different writing process elements appropriately to communicate with different audiences for a variety of purposes (NCTE/IRA Standard #5); and students will apply knowledge of language structure, language conventions, media techniques, figurative language, and genre to create, critique, and discuss print and nonprint texts (NCTE/IRA Standard #6).

Performance Tasks

- Immerse students with poetry from various collections appropriate for young adults. Notable collections include anthologies by Mel Glenn, Lori Carlson, Paul Janeczko, Paul Fleischman, and others. See the Bibliography at the end of this chapter for a variety of poetry collections. Require students to create a poetry portfolio on a theme of their choice. Read Out of the Dust by Karen Hesse for a novel written in free-verse to advance the plot. Ask students to write another diary entry for Billie Jo in Hesse's style. See *Poetry Anthology Requirements* on page 133.

Poetry Anthology Requirements

Create Your Own Poetry Anthology

For this assignment, you will collect published poems on a theme of your choice, illustrating some of them, explicating one of them, and writing three poems of your own.

Procedure

- ♦ Choose a theme for your anthology, such as nature, friendship, sports, joy, fantasy, people.
- ♦ Collect 10 poems from the poetry anthologies on the booklist that develop your theme.
- ♦ Illustrate at least five of these poems using photos, original drawings, pictures from magazines, cartoons, etc.
- ♦ Include five newspaper or magazine articles related to your theme.
- ♦ Write three original poems on your theme using any form of poetry you prefer, such as free verse, rhymed verse, sonnet, cinquain, haiku, concrete, senses, diamante, found, poem in two voices, etc.
- ♦ Explicate one poem, either yours or one in your collection.
- ♦ Vocabulary page with a minimum of 10 new words and their definitions found in your collection of poems.
- ♦ Write an introduction to your anthology that introduces the reader to your theme and that summarizes your experiences compiling the anthology.
- ♦ Write a self-reflection. What did you learn from this project? What goals will you set for the next time you write poetry?
- ♦ "Works Cited" page for poetry selections, newspaper and magazine articles.

- To encourage students to play with words and experiment with various forms of poetry refer to *Getting the Knack: 20 Poetry Writing Exercises* by Dunning and Stafford (1992). They include poetry writing exercises such as found and headline poems, letter poems, acrostic and recipe poems, as well as many other forms. Another reference is *Enriching Our Lives: Poetry Lessons for Adult Literacy Teachers and Tutors* by Kazemek and Rigg (1995). Each of the eight chapters includes a sample lesson for a different type of poetry, such as poetry in conversation, form poetry, music as narrative poetry. Students should include their best three poems in their *Poetry Anthology*. See *Poetry Assessment Rubric*.
- Prepare students for the Coffee House by practicing various types of poems that have strong rhythms and repetitions that lend themselves to oral performance. Encourage students to add appropriate physical movements such as hand clapping, arm waving, foot tapping, table drumming. Students may want to add a musical background to their performances. Read and perform Fleischman's "Water Boatmen" from *Joyful Noise: Poems for Two Voices* (1988). This poem helps us see and feel two things at once—insects swimming under water and people rowing a racing shell on top.
- Ask students to select and present their favorite original poem from the poetry anthology. Award bonus points to poets who memorize their work of art. See *Assessment Rubric for Coffee House Oral Performance* on page 136.
- Schedule the Coffee House during regularly scheduled class time, unless you prefer to stage the performance at a local restaurant or coffee house and include parents, friends, and community members. Consider a fundraiser or grant to bring in a guest poet to model reading poetry for the students and provide feedback for the poems they perform. Provide a microphone and speakers for students who might need musical accompaniment. You might even invite an acoustic guitar player to play background music during the performance. Provide refreshments if the performance is held in the classroom. At the end, ask the audience to nominate and vote for their favorite student poets by a show of hands. Award prizes of books, T-shirts, and cookies to the top 10 students.

Poetry Assessment Rubric

	Publishable (4 points)	*Excellent (3 points)*	*Revised (2 points)*	*Draft Stage (1 point)*	*Score*
Message/ Meaning (mood, key point, image)	Thought-provoking; evoked an emotional response; passionate; evocative, empathetic; experiential; intense. The message is fully developed.	Clear message focused on a central idea, feeling or experience and engages the reader. There are minor lapses in development and/or clarity.	The poem lacks focus and does not completely engage the reader. The message needs more development and clarity.	The poem does not focus on a central thought and does not engage the reader. The message is confusing.	___×5=___ (20 points)
Form/ Organization	The form conveys the meaning; form provides appropriate, even unique, structure for the main idea. Poem makes effective use of poetic elements such as sound devices, rhyme, refrain, onomatopoeia, and meter.	The form is adequate in providing structure for the main idea. Poem makes use of limited poetic elements such as sound devices, rhyme, refrain, and meter.	The form is not adequate in providing structure for the main idea. It uses prose or sentence format. Little evidence of poetic elements.	Form is lacking poetic format. There is no use of any standard poetic elements.	___×5=___ (20 points)
Imagery, Use of Figurative Language, Symbolism	Radiant images, sensuous, vivid use of language and poetic devices. Unexpected comparisons are made.	Clear attempt to create effective images using poetic language and devices. Some use of metaphor and/or simile.	Limited command of poetic language and devices. There are errors in the use of language.	Reads like prose instead of poetry. Denotative rather than connotative language; does not show command of basic literary devices. There are major errors in the use of language.	___×4=___ (12 points)
Illustrations	The illustration captures the readers attention and enhances the meaning of the poem.	The illustration captures the reader's attention, but does not enhance the meaning of the poem.	Artistic modes are not well-developed and detract from the meaning of the poem.	The illustrations lack technical competence and detract from the message.	___×3=___ (12 points)

Assessment Rubric for Coffee House Oral Performance

	Exemplary (4)	*Proficient (3)*	*Developing (2)*	*Beginning (1)*	*Score*
Stage Presence	Student uses gestures and mannerisms to enhance the delivery and does so naturally and effectively	Student uses some gestures and mannerisms to enhance the delivery	Student uses few or inappropriate mannerisms that detract from the presentation	Student uses distracting behaviors that impairs the message	__×7=__ (28 points)
Presentation of Poems	Student enunciates clearly, uses expression, and is clearly heard by the audience	Student speaks clearly, but with little expression	Student does not enunciate and is difficult to hear	The pronunciation and enunciation are so unclear that one cannot understand the poem	__×7=__ (28 points)
Visual, Physical or Musical Effects (Optional)	Student uses visuals in unique and interesting ways to enhance the message	Student uses effective visuals to support the point of the message	Few visuals are used or visuals are used ineffectively	No visuals are used to support the message	__×5=__ (20 points)
Eye Contact (Bonus, if memorized)	Student establishes good eye contact with audience and makes audience members feel as if they are being spoken to	During most of the presentation, student maintains eye contact with audience	Student establishes minimum eye contact with audience	Student establishes no eye contact with the audience	__×3=__ (12 points)
Audience Response	Student communicates meaning to peers; holds audience attention	Student communicates some meaning to peers, but the poem is too long or difficult to follow	Student does not communicate meaningfully and does not hold audience attention	The presentation is ineffective and boring	__×3=__ (12 points)

Bibliography

Adoff, A. (1995). *Slow dance heart break blues.* New York: Lothrop, Lee & Shepard Books.

Carson, J. (1989). *Stories I ain't told nobody yet.* New York: Orchard Books.

Carlson, L. M. (1994). *Cool salsa: Bilingual poems on growing up Latino in the United States.* New York: Henry Holt.

Cronyn, G. W. (Ed.). (1991). *American Indian poetry: An anthology of songs and chants.* New York: Ballantine Books.

Cullinan, B. E. (Ed.). (1996). *A jar of tiny stars: Poems by NCTE award-winning poets.* New York: Boyds Mills Press.

DeSoto, G. (1992). *A fire in my hands.* New York: Scholastic.

Dunning, S., Lueders, E., & Smith, H. (1967). *Reflections on a gift of watermelon pickle… and other modern verse.* New York: Lothrop, Lee & Shepard Books.

Dunning, S., & Stafford, W. (1992). *Getting the knack: 20 poetry writing exercises.* Urbana, IL: National Council of Teachers of English.

Fleischman, P. (1988). *Joyful noise: Poems for two voices.* New York: HarperTrophy.

Fleischman, P. (1989). *I am phoenix: Poems for two voices.* New York: HarperTrophy.

Fox, J. (1997). *Poetic medicine.* New York: Putnam.

Glenn, M. (1982). *Class dismissed! High school poems.* New York: Clarion.

Glenn, M. (1986). *Class dismissed II: More high school poems.* New York: Houghton Mifflin.

Glenn, M. (1988). *Back to class: Poems by Mel Glenn.* New York: Clarion.

Glenn, M. (1991). *My friend's got this problem, Mr. Candler.* New York: Clarion.

Glenn, M. (1997). *The taking of room 114: A hostage drama in poems.* New York: Lodestar Books.

Glenn, M. (1996). *Who killed Mr. Chippendale?: A mystery in poems.* New York: Lodestar Books.

Hesse, K. (1997). *Out of the dust.* New York, Scholastic.

Janeczko, P. B. (Ed.). (1993). *Looking for your name: A collection of contemporary poems.* New York: Orchard Books.

Janeczko, P. B. (1990). *The place my words are looking for.* New York: Bradbury Press.

Janeczko, P. B. (1985). *Pocket poems: Selected for a journey.* New York: Bradbury Press.

Janeczko, P. B. (1991). *Preposterous: Poems of youth.* New York: Orchard Books.

Janeczko, P. B. (1984). *Poetry from A to Z: A guide for young writers.* New York: Bradbury Press.

Janeczko, P. B. (1991). *Poetspeak: In their work, about their work.* New York: Collier Books.

Janeczko, P. B. (1984). *Strings: A gathering of family poems.* New York: Bradbury.

Kazemek, F. E., & Rigg, P. (1995). *Enriching our lives: Poetry lessons for adult literacy teachers and tutors.* Newark, DE: International Reading Association.

13

THE SOCRATIC SEMINAR

Susan Malaska

Let him that would move the world first move himself.

Socrates

Overview of Authentic Learning

Socratic questioning is both an innovative and ancient technique that requires students and teachers to assemble in an open forum to discuss the comprehension of a work and the important concepts proposed by that work. In ancient Greek times, Socrates raised thoughtful questions without proposing to have all of the answers himself. Through discussion and dialogue, he hoped to lead students to discover truth and self-knowledge. Socratic Seminars may have ancient traditions, but they always remain innovative since seminars rely on a modern day interpretation of a text and connect students to literary themes in the light of their current experiences.

However, this is not to say that a Socratic Seminar is just a "bull session" without purpose or structure. On the contrary, the seminar requires much thoughtful preplanning on the part of the teacher, and serious, individually assessed work from the student. In fact, the seminar can be structured to fulfill any of the reading or writing competencies as well as most of the listening or speaking competencies in any district's course of study. The teacher's work consists of assigning a worthy topic, problem, or reading; preparing sequenced, thought-provoking questions; and then skillfully provoking students to think, to exchange ideas, and to respect others' opinions. The teacher carefully backs out of the conversation and takes the role of observer, evaluator and often coach to keep the discussion moving and on track. "It is a unique alternative to traditional class discussion because in seminar students speak 97% of the time" (Canady and Rettig, 1996, p.43).

A seminar supported by a thoughtful structure of questions from the teacher, imbedded with purpose, respect, and trust, allows students not only intellectual freedom to explore ideas, but the structure allows direct teacher observation of reading, writing and thinking skills. Overwhelmingly, the majority of my students say they are surprised and delighted that there are so many opinions and ways to look at an idea, but they also reflect on increased comprehension of a work.

The larger goal of a seminar is to extend students' thinking by exploring common themes and original viewpoints. The more complex goals can range from assessing students' abilities to clearly summarize an idea, to assessing implied meanings in a passage. The rewards in seminar are varied and often surprising. The teacher learns as much as, or more than, the students do. Seminar provides a concrete way to measure growth in reading, writing, speaking, and listening while watching students take charge of their own learning.

The following seminar is designed to follow the reading of Night, by Elie Weisel, but works as well for other nonfiction novels, fictional novels, short stories, articles, movies, newspaper editorials, problem-solving situations, etc.

NCTE/IRA Standards

In the Socratic Seminar assessment, students will:

- read various works about the Holocaust. In doing so, students will read a wide range of print and nonprint texts to build an understanding of texts, of themselves, and of the cultures of the U.S. and the world; to acquire new information; to respond to the needs and demands of society and the workplace (NCTE/IRA Standard #1), and they will read a wide range of literature from many periods in many genres to build an understanding of the many dimensions of human experience (NCTE/IRA Standard #2).
- keep a reader response log reflecting on their insights into the text. In doing so, students will apply a wide range of strategies to comprehend, interpret, evaluate, and appreciate texts. They draw on their prior experience, their interactions with other readers and writers, their knowledge of word meaning and of other texts, their word identification strategies, and their understanding of textual features (NCTE/IRA Standard #3); they will employ a wide range of strategies as they write and use different writing process elements appropriately to communicate with different audiences for a variety of purposes (NCTE/IRA Standard #5); and they will apply knowledge of language structure, language conventions (e.g., spelling and punctuation), media techniques, figurative language, and genre to create, critique, and discuss print and nonprint texts (NCTE/IRA Standard #6).

- take responsibility for their part in the seminar discussion. In doing so, students will participate as knowledgeable, reflective, creative, and critical members of a variety of literacy communities (NCTE/IRA Standard #11) and students will use spoken, written, and visual language to accomplish their own purposes (NCTE/IRA Standard #12).

Performance Tasks

Performance Tasks for Preseminar Activities

- Students make one-minute vocabulary reports on Holocaust terms, places, and concepts. Assign each student a vocabulary word and ask them to find out the meaning of the word and design a visual for their presentation. After the presentations, post the visuals around the room as a reminder of the word throughout the reading and discussion of Night. See Holocaust Vocabulary taken from the Schools of California Online Resources for Education's SCORE: Teacher Guide for Night found at http://www.sdcoe.d12.ca.us/score/night/nighttg.html The definitions for the words can also be found at this site.

Holocaust Vocabulary

Anti-Semitism
Aryan
Assimilation
Auschwitz
Babi Yar
Bar Mitzvah
Belzec
Bergen-Belsen
B'richa
British White Paper of 1939
Buchenwald
Bund
Cantor
Chelmno
Communism
Concentration Camp
Dachau
Death Camp
Death Marches
Dehumanization
Desecrating the Host
Diaspora
Displacement
DP
Displaced Persons Act of 1948
Drancy
Einsatzgruppen
Euthanasia
Final Solution
Fuhrer
Gas Chambers
Genocide
Gestapo
Ghettos
Gypsies
Holocaust
Homophobia
Jehovah's Witnesses
Judaism
Kippah
Koreczak, Dr. Janusz
Kristallnacht
Majdanek
Marranos
Muselmann
Nuremberg Laws
Operation Reinhard
Pale of Settlement
Pogrom
Prejudice
Propaganda
Ravensbruck
Resettlement
Revisionists
Scapegoat
SD
Hannah Sennesh
Shoah
Sobibor
Social Darwinism
Sonderkommando
SS
Star of David
Stereotype
Theresienstadt
Treblinka
Umschlagplatz
Underground
Raoul Wallenberg
Wannsee Conference
Warsaw Ghetto
Zyklon B

- Students read Elie Weisel's Night. Students read and write daily responses to the text reflecting on observations, questions, or meaningful quotes. Students should be encouraged to make connections with their own personal experiences and make connections with other texts, concepts, or events. See *Response Journal Prompts* on page 143 contributed by Rita Elavsky, Roberts Junior High School, Cuyahoga Falls, Ohio.
- On the announced day of the seminar, the classroom is arranged in a way to promote discussion (a circle usually works the best). In a large class, an inner and outer circle work better. The students in the outer circle have assigned duties such as taking notes, collecting student quotes, observing speaking patterns, or giving feedback to the inner circle after the discussion is over.
- A rubric detailing how the student will be assessed is distributed and discussed. The teacher prepares a chart with all of the students' names to keep track of numbers and types of responses. The chart includes frequency of comments, opinions, questions, text connections, classmate connections, speaking skills, listening skills and conduct.

Performance Tasks for the Seminar Discussion

- Begin with an Opening Question. It needs to be general and one that encourages

students to get into the text, its themes and main ideas:

What is Night really about?

The discussion often begins with a "round robin" approach to make sure everyone gets an opportunity to make an initial comment. I copy down the comments or often have a student list them on the board. These comments can automatically contribute to the Core Questions, as the teacher becomes more skilled.

- Continue with three to five Core Questions. These questions should be designed to promote discussion of factual information and personal reactions. They need to be more specific and are often "how" or "why" questions. The most desired answers are those that reflect student opinion backed up with factual detail from the text.
  - What was it like in a Nazi concentration camp?
  - How could the Nazi soldiers do what they did?
  - Why did Elie survive when others did not?
  - How did Elie change through his years in the camps?
  - What kinds of examples of self-survival were in the book?

Response Journal Prompts

- Share a personal experience you have had with racism, sexism, anti-Semitism, or other discrimination or consider what groups have been victims of prejudice over the course of history. What happened as a result of this prejudice? Discuss how prejudice and discrimination are not only harmful to the victim but also those who practice them.
- Imagine that your home or business was attacked during the Kristallnacht. You must clean up the mess. Place some of the articles you find in an imaginary box and list why you consider them significant. For example, broken glass symbolizes the violence of discrimination; a broken toy demonstrates how children are innocent victims of hatred; a broken table leg signifies the table where the family gathered for evening meals; a Star of David symbolizes the religious devotion that the Nazis found so offensive.
- You are faced with the school bully, who wants to take away your lunch. He outweighs you by 50 pounds. Do you fight? Fight back "dirty"? Run? Get help? Tell your parents? Explain what you do and what you think should come from the incident.
- Imagine yourself being transported several hundred miles in a filled-to-capacity cattle car or boxcar, with no idea of your destination. The trip could last a few hours or a number of days. Describe in vivid detail what the experience might be like. Describe the sights, the sounds, and the smells, as well as the diverse group of people with whom you are traveling.
- In discussing the Holocaust, one survivor, Luba Frederick, said, "To die was easy." Based on what you know from your study of the Holocaust, explain her statement.
- Could events similar to those that led to the Holocaust occur today in the United States? Could rights, for example, be taken away from a specific group of people, and could public attitude towards a single group be completely altered? Explain your answer.
- Is there any value in continuing the search for, and the prosecution of, Nazi war criminals, most of whom are feeble old men, or should they be left alone to die in peace?
- How does selecting and wearing shoes or clothing relate to your personal identity? Can you retain your personal identity when forced to abandon your personal belongings? What does the document itemizing the victims' belongings tell you about the person who wrote it? What questions do you have about how these records were created and kept?
- How would you define the word hope? What part does hope play in survival? What role did hope play in Elie's life and survival?

At this point, students call on one another and the teacher's job is to observe and chart, not contribute to the conversation unless it goes off the subject or contains erroneous information. The teacher might interject with: "Are you saying that...?"

"Where in the text do you find support for what was just said?" "Does anyone want to challenge what was just said?"

- ♦ The Closing Questions establish relevance and connect students to the real world. This is often the most animated and rewarding part of the seminar.
  - What kinds of prejudice exist at Shelby High School?
  - Why are some students treated differently than others?
  - What can you do to lessen the cruelties you witness at Shelby High School?

Performance Tasks for Post-Seminar Activities

The seminar can lead to a variety of postseminar tasks that enrich and extend learning, as well as self-reflection.

- ♦ Activities can include research, a piece of art, a letter to an author or agency, a paragraph designed to discuss bias or propaganda, predictions, and/or the author's purpose. Here are a few suggestions for starters:
  - Find an article on the Internet about Elie Weisel today and discuss it.
  - Why do you think Weisel wrote this book? In what ways has his life since been molded by the horrific events of the Holocaust?
  - Read and comment on Weisel's Nobel Peace Prize Acceptance Speech.
  - Investigate the life of David Olère, a Holocaust survivor and artist.
  - How were the propaganda techniques of scapegoating and stereotyping used by the Third Reich to condone and even encourage behavior that most German citizens would have considered abhorrent?
  - What was the "Final Solution" and how were its objectives carried out in places like Jewish ghettos and concentration camps like Auschwitz, Berkinau and Dachau?
  - Read, I Never Saw Another Butterfly, and write a letter to one of the children commenting on their poetry and art.
  - Many laws that limited the freedom of Jewish citizens were written in Germany when the Nazis came to power. Select one of those laws and examine the reason for the law, its impact on Jew-

ish and non-Jewish citizens, and its relationship to other events taking place in Germany at the time.

- Investigate the reasons why specific groups became victims of the Nazis, including children, Gypsies, Blacks, Jehovah's Witnesses, the handicapped, homosexuals, and others, and investigate the reasons for their respective treatment.

♦ Guide students self-reflection by providing questions such as these:
- What did you learn during this seminar and project experience?
- What did you do well?
- What are you confused about?
- What do you need help with?
- What do you want to know more about?
- What did you learn that you can transfer into the "real world"?
- What will you do differently next time?

♦ In order to inform students of their final seminar grades, use a previously introduced rubric that includes points for prework, participation and postwork. See *Socratic Seminar Assessment Rubric* on page 146.

Socratic Seminar Assessment Rubric

	Exemplary (4)	*Proficient (3)*	*Apprentice (2)*	*Beginner (1)*	*Score*
Pre-Seminar Activities					
Vocabulary Report	Creative use of color and design in the visual; complete, accurate definition with examples; invited audience to learn more	Uses color and artful design in the visual; good definition of the concept; interesting presentation	Limited use of color and design in the visual; ambiguous definition lacking sufficient details; lacking energy in the presentation	Poor planning for layout, color, and design in the visual; inadequate or incorrect definition; monotone or bored sounding presentation	__×3=__ (12 points)
Reader Response Log	Entries include detailed, even moving reflection of the thoughts, feelings, and associations made during the reading experience	Entries include reflection on the thoughts, feelings and associations made reading experience and are personal, though not necessarily moving	Entries lack details and are more descriptive than reflective of the reading experience	The entries are sketchy and ambiguous demonstrating limited control of language and thought	__×4=__ (16 points)
Seminar Discussion					
Contributions	Contributions are insightful; gives illustrations, examples to support and clarify ideas	Makes useful contributions to move the discussion forward; expresses opinions freely	Makes limited contributions to the discussion that are more opinion than research based	Unprepared; does not contribute to the discussion	__×6=__ (24 points)
Questions	Asks higher level questions that generalize, synthesize, and evaluate	Asks some higher level questions; asks application questions	Asks questions that require lower level responses, such as memory and recall	Does not ask questions	__×5=__ (20 points)
Textual Connections	Provides accurate proof, data and textual evidence for contributions	Provides some textual evidence to support ideas	Makes surface level references to the text	Never refers to text or research; relies on limited personal opinion	__×3=__ (12 points)
Classmate Connections	Encourages other students to participate; extends classmate's ideas	Acknowledges classmate's ideas before presenting own ideas	Sometimes acknowledges classmate's ideas before presenting own ideas	Never acknowledges classmate's contributions before presenting own ideas	__×3=__ (12 points)

Speaking Skills	Uses rich vocabulary; is enthusiastic; clear and audible	Uses appropriate vocabulary; usually interesting, clear language; audible	Good use of language; generally clear articulation; generally audible	Demonstrates limited language development; monotonous and/or monotone	__×2=__ (8 points)
Listening Skills	Uses active listening skills such as paraphrasing, clarifying, and questioning classmates' contributions	Uses some active listening skills; jumps into the discussion impulsively	Does not use active listening skills; interrupts to make a point	Inattentive; daydreaming; absentminded; loses track	__×2=__ (8 points)
Conduct	Polite, respectful, attentive, focused	Usually polite, respectful and attentive	Sometimes polite, respectful and attentive	Off task, disruptive behavior	__×2=__ (8 points)
Postseminar Activities					
Self-Reflection	Clearly articulates and elaborates on learning, progress, and goals	Good ability to reflect on learning, progress, and goals	Describes learning experiences rather than reflecting on them	No attempt to describe or reflect on learning experiences, progress, and/or goals	__×1=__ (4 points)
Project	Project is a creative, original expression; excellent application of critical analysis, interpretation and judgment	Substantially completed project; some evidence of analysis, interpretation and critical judgment	Minimally completed project; little use of analysis, interpretation or critical judgment	Shows little evidence of attempting the project; no evidence of analysis, interpretation or judgment	__×5=__ (20 points)
				Total Points (144 possible) Grade Scale A = B = C = D =	

Bibliography

Canady, R. L., & Rettig, M. D. (1996). *Teaching in the block: Strategies for engaging active learners.* Larchmont, NY: Eye On Education.

Schools of California Online Resources for Education. SCORE: *Teacher guide—Night* by Elie Weisel [Online]. Available: http://www.sdcoe.d12.ca.us/score/night/nighttg.html.

Volavkova, H. (Ed.). (1993). *I never saw another butterfly: Children's drawings and poems from Terezin concentration camps 1942–1944.* New York: Pantheon Books.

Weisel, E. (1982). *Night.* New York: Bantam Books.

14

STORYTELLING/ MASK PROJECT

Janie Reinart

...I was asking him to look at his life, pick out what mattered, and commit his thinking about it to paper—to become an originator of knowledge.

Tom Romano

Overview of Authentic Learning

How do students tell their own stories, and learn more about themselves and others? Creating a 3-D mask helps students to find their voice, share their original story in a presentation, and explore their identity inside and out. In a one- to two-page, typed essay, students describe themselves using a mask to illustrate their physical, mental, emotional, and moral self. Because identity is a combination of how you see yourself and how you wish to be seen by others, students have an opportunity to answer the question "who are you really?" The outside of the mask represents how others see them, whereas the inside of the mask shows who they really are. Students have several weeks to write and create their masks and essays. Presentation time depends on the size of the class. A teacher-created rubric assesses the mask, the essay, and the presentation. The rubric has a self-evaluation component worth five points. Students write on Post-it notes after their presentation to defend the number grade they give their project. Start the project by reading poems or listening to music that shows society wearing masks.

NCTE/IRA Standards

In this assessment, students will:

- read and respond to a poem and song lyrics. In doing so, students will read a wide range of print and nonprint texts to build an understanding of texts, of themselves, and of the cultures of the U.S. and the world; to acquire new information; to respond to the needs and demands of society and the workplace; and for personal fulfillment (NCTE/IRA Standard #1).
- write a class poem and a personal narrative. In doing so, students will employ a wide range of strategies as they write and use different writing process elements appropriately to communicate with different audiences for a variety of purposes (NCTE/IRA Standard #5); and students will apply knowledge of language structure, language conventions (e.g., spelling and punctuation), media techniques, figurative language, and genre to create, critique, and discuss print and nonprint texts (NCTE/IRA Standard #6).
- create a mask to represent the ones they wear. In doing so, students will adjust their use of spoken, written, and visual language to communicate effectively with a variety of audiences and for different purposes (NCTE/IRA Standard #4); students will participate as knowledgeable, reflective, creative, and critical members of a variety of literacy communities (NCTE/IRA Standard #11); and students will use spoken, written, and visual language to accomplish their own purposes (NCTE/IRA Standard #12).
- tell their stories and share their masks with the class. In doing so, students will participate as knowledgeable, reflective, creative, and critical members of a variety of literacy communities (NCTE/IRA Standard #11), and students will use spoken, written, and visual language to accomplish their own purposes (NCTE/IRA Standard #12).

Materials

Students need to use their imagination in making these 3-D masks. Some examples for materials are:

- Legos/building blocks
- Paint/paper mache
- Collage/construction paper
- Hats/miniature items
- Cookies with decorations and words
- Photos and Styrofoam

Students also need post it notes for the self-evaluation segment of the project.

Performance Tasks

- Read the poem, We Wear the Mask, by Paul Laurence Dunbar.

  Discussion: What is the author saying to you? What are some of the masks we wear? What is the cost of wearing a mask? What do we want to hide? What are we afraid to show?
- Listen to the song, *Facade,* from the musical *Jekyll & Hyde.*
  - As students listen to the song, ask them to write down words or phrases that appeal to them in their journals
  - When the song is finished, ask students to write a favorite phrase, word, or repetition of words on strips of different colored paper (any size that can be seen from seat).
  - One at a time, ask students to arrange strips of paper with phrases on the board to create a class poem. (Use masking tape to stick the strips to the board.)
  - Vote on a title for the poem.

Mask Making

Describe and create a mask that shows your identity. Identity is a combination of how you see yourself and how you wish to be seen by others. Who are you really? See *Assessment Rubric for Storytelling/Mask Project* on page 153.

- In a one- to two-page, typed essay, describe your mask, telling us about your physical, mental, emotional, and moral self. Tell us who you really are.
- Create a mask. On the outside of the mask use colors, symbols, designs, and decorations to reveal to us the "self" that you show teachers, parents, and friends. On the inside of the mask use words, pictures, symbols, and decorations to show the real you and what is in your heart. You will present your essay and mask to the class.
- See on page 152 *Eric's Story,* Eric Stevens' explanation of his mask that was made with red Lego blocks with objects and drawings attached to the shape.
- See Figure 14.1 for Sara's mask. Here's what Sara said about the project in her inkshed: "*I thought the project was an interesting way to let people know how I actually feel about myself and how what they think of me compares to what I think of myself… I really liked doing this and it made me learn about myself more. I think it's a very healthy way for students to let out their emotions and I think more students should be doing this project because it actually made me feel pretty good about myself. Thank you for asking me to do a mask.*"
- Share your mask and tell your story to the class.

Eric's Story

Eric Stevens
4–5 period
5–3–99

The outside of my mask, for the most part is exactly identical to the inside. There are a few differences, though. Sometimes I feel one way and act another way which makes me phony at times.

The outside of my mask had a big fat smile. This smile represents the joy and happiness that I show on the outside. I also have a clown in the middle of the mask. This clown figurine represents the silly, uncontrollable joy that I portray to others. Clowns tend to be silly, bizarre, funny, and weird all at the same time which are all the words that describe me. Keep in mind I'm not like this all the time.

I believe the goofy, funny aspect of my personality outweighs my serious side, though I can be serious. When there's turbulence in my life, such as a death, or when it comes to people's problems, or just people in general, I get serious. Also, my schoolwork is a vital thing that's important to me which converts me into my serious mode. Basically, things that are important to me keep me away from taking those vital things lightly and turns my fun, natural self down to zero. The small pencil on the top of my mask symbolizes my seriousness. The size of my pencil symbolizes the frequency of my seriousness.

The boat drawing on the side of my mask signifies the military background that I've grown up with. I have three older brothers who are all currently serving in the Navy and a father who was a submarine dude during the Vietnam war. Naturally, I have grown accustomed to the ways of the Navy and have also grown an interest in the Navy. I have picked up some military characteristics throughout my eighteen years on this planet and thus merges with the rest of my character. Being a sailor and my basic knowledge and experience on the water is a big chunk of me. My desire to be by the water will have an influence on my future. I want to become a Naval Officer.

Figure 14.1 Sara's Mask

Assessment Rubric for Storytelling/Mask Project

	Exemplary (4)	*Proficient (3)*	*Developing (2)*	*Beginning (1)*	*Score*
3-D Mask	Mask exhibits superior understanding and application of design, texture, color, shapes, and symbols to express your personality	Mask exhibits good ability to apply principles of design, texture, color, shapes, and symbols to express your personality	Mask exhibits little understanding of the principles of design, texture, color, shapes, and symbols to express your personality	Mask exhibits no understanding of the principles of design, texture, color, shapes, and symbols to express your personality	__×7=__ (28 points)
Essay Content	The essay is highly coherent; free-flowing ideas; well-organized with many personal details	The essay is coherent, ideas flow well; is organized and has personal description and detail	The essay lacks cohesiveness; lacks thoughtful organization; minimal use of detail	The essay lacks cohesiveness; writing is sporadic; very little development of ideas; no details	__×7=___ (28 points)
Essay Conventions	The author uses correct sentence structure, grammar, punctuation and spelling	The author uses mostly correct sentence structure, grammar, punctuation and spelling	The author makes errors in sentence structure, grammar, punctuation and spelling	The author makes many errors in sentence structure, grammar, punctuation and spelling	__×2=__ (8 points)
Presentation	The speaker delivers the message in a lively, enthusiastic fashion	The speaker delivers the message in an easy-to-listen-to manner	The speaker's message is difficult to hear and understand	The speaker mumbles words and is inaudible	__×5=__ (20 points)
Self-Evaluation	Excellent understanding of personal growth, learning, and goal setting	Good understanding of personal growth, learning, and goal setting	Minimal understanding of personal growth, learning, and goal setting	No understanding of personal growth, learning, and goal setting	__×4=__ (16 points)

Total Points (100 possible)

Grade Scale
A =
B =
C =
D =

Bibliography

Dunbar, P. L. (April 2000). *We wear the mask* [Online]. Available: http://www.library.utoronto.ca/utel/rp/poems/dunbarp5.html.

"Façade." CD of Broadway musical *Jekyll and Hyde* [Online]. Original Broadway cast recording. Performers include Robert Cuccioli, Linda Eder, and Christiane Noll. (No date given.)

(April 2000). *Jekyll and Hyde* [Online]. Available: http://jekyll-hyde.com/welcome-b.html.

Romano, T. (1995). *Writing with passion.* Portsmouth, NH: Heinemann.

15

TOLERANCE TALENT SHOW AND ART EXHIBITION

Joyce Rowland

Chance favors only the prepared mind.

Louis Pasteur

Overview of Authentic Learning

The culminating activity for this unit is for students to demonstrate their understanding of tolerance and respect for others by participating in either the Talent Show or the Art Exhibition to be held for parents, administrators, and classmates. For classes studying this unit in late October, students might imitate the "Maycomb County: Ad Astra Per Aspera" original pageant composed by Mrs. Grace Merriweather in *To Kill a Mockingbird* (1960, p. 290). Otherwise, students should choose from the following list of projects according to their intelligence type to create a skit, monologue, musical rendition, or dance for the Talent Show that conveys what tolerance means to them. Artwork and sculptures will be displayed at the big event, as well. Another table will be needed for any food that the students might prepare (performance assessment is much more rewarding for the class than a test!). Schedule the auditorium for a convenient date and arrange for display tables for the artwork, maps, time lines, and board games. Give students time to prepare their performances and then schedule their acts accordingly. Students may work collaboratively in groups to strengthen their interpersonal intelligence or individually to strengthen their intrapersonal intelligence.

NCTE/IRA Standards

In this assessment, students will:

- create a project (skit, monologue, musical rendition, artifact) to present at the Talent Show and Art Exhibition on Tolerance. In doing so, students will conduct research on issues and interests by generating ideas and questions, and by posing problems. They will gather, evaluate, and synthesize data from a variety of sources to communicate their discoveries in ways that suit their purpose and audience (NCTE/IRA Standard #7); students will use a variety of technological and informational resources to gather and synthesize information and to create and communicate knowledge (NCTE/IRA Standard #8); students will participate as knowledgeable, reflective, creative, and critical members of a variety of literacy communities (NCTE/IRA Standard #11); and students will use spoken, written, and visual language to accomplish their own purposes (NCTE/IRA Standard #12).

Performance Tasks

Students choose from the following list of activities depending on their preference of multiple intelligences:

Verbal/Linguistic Intelligence Activities

- Create a poem for each main character that acts as a character sketch. Write an analysis of one of your poems. Choose at least one poem to perform.
- Write a series of newspaper articles covering the trial of Tom Robinson, including the article after his death. Prepare a display for the Art Exhibition.
- Prepare a debate between a 1933 Southern lawyer and a modern day feminist arguing why women should or should not be allowed to serve on a jury.
- Create a family tree for Jem and Scout, and then create a family tree for yourself, going back at least five generations for each. Display the results at the Art Exhibition.
- Write a comparison and contrast skit of Tom Robinson's trial, comparing/contrasting it to O.J. Simpson's trial.
- Write a comparison and contrast skit of Tom Robinson, comparing/contrasting him to an American soldier in Vietnam.
- Make a 60-second radio endorsement for Atticus Finch for President of the United States.
- Write an essay and prepare a skit comparing Atticus Finch to Sherlock Holmes.

- Make a Jeopardy board based on the book *To Kill a Mockingbird*. Let the class play the game with you as an emcee.
- Keep a personal journal for Boo Radley. Include 10 entries and prepare a monologue.
- Research the laws governing women on juries and make a presentation on how, when, where, and why they changed.
- Write a new law for Maycomb. Justify its need at the Talent Show.

Logical/Mathematical Intelligence Activities

- Make a map of the Finch's neighborhood, labeling anything you can possibly label, including where specific incidents happened. You may even include the town if you want. Display at the Art Exhibition.
- Create for display a time line of the African American's plight in America.
- Create a time line for each of the plots in the story (Boo Radley and Tom Robinson) on a computer using Hyperstudio or a similar slide show program.
- Rank and define the social classes found in Maycomb in the 1930s. Plot them on a chart, including names and classifications of the characters in the book, for display.
- Analyze in a panel discussion the cause and effect of the verdict given in the Tom Robinson trial.
- Gather recipes for, and make, a Southern-style meal to share with the class. (I certainly hope you include grits!) Provide copies of all recipes used, including the calculations needed to increase the recipes to make enough for the entire class. Make the meal and serve it.
- Make Atticus' outline for his case in Tom Robinson's behalf; perform as a monologue.
- Compare in graphic format the social class system in Maycomb to that of Feudal Europe.
- Make and explain a flow chart of the action of the story for the Art Exhibition.

Bodily/Kinesthetic Intelligence Activities

- Do a frieze display of at least eight scenes from the book.
- First demonstrate the game of tire rolling with a stick. Then demonstrate how Scout rolled in the tire to Boo's house. If possible, show a film of what Scout would see as she rolled in the tire. Find and demonstrate other games that children of the Depression may have played.

- Act out a scene from the book.
- Dress as, and take on, the character of one of the people in the book. Visit the Talent Show and talk to us in the dialect of that person. Be prepared to answer our questions.
- Dress as, and take on, the character of a famous black leader: a musician, dancer, poet, athlete, politician, or teacher. Prepare a monologue for the Talent Show.
- Make up a dance that demonstrates the feelings of Tom Robinson before, during, and after the trial.
- Demonstrate whittling. Whittle figurines of Jem and Scout either out of wood or soap.
- Set up a scavenger hunt to locate the things that Boo gave to Jem and Scout.
- Demonstrate the best way to create a flower garden. Include information on camellias.
- Choreograph and perform a dance representing the attack scene when Scout is in her ham costume.

Visual/Spatial Intelligence Activities

- Make a map of the neighborhood and town using Virtual Reality software. Take us on a tour.
- Create a photo story representation of the story.
- Make a collage representing the themes of the book.
- Design and paint a mural showing Scout growing up.
- Create a chart in which you graph the violence throughout the book.
- Draw a storyboard representing the scenes in the book.
- Collect and make a museum display of articles from the Depression.
- Mindscaping: "Images of Tolerance."

Musical/Rhythmic Intelligence Activities

- Collect, play, and explain the significance of music from the Depression.
- Write and perform the "Ballad of Tom Robinson."
- Perform at least 10 sound effects from the book and let the audience figure out what they are.
- Create and perform a rap about the book.
- Line a hymn to the class as Calpurnia's son might have done.
- Create an audiotape of music clips representing a chronology of scenes from the book.
- Perform music from the Civil Rights movement.

Assessment Rubric for the Talent Show

	Exemplary (4)	*Proficient (3)*	*Developing (2)*	*Beginning (1)*	*Score*
Topic	Student addresses the topic creatively and provides vivid examples to illustrate the main points in the project	Student addresses the topic clearly, but provides few examples to illustrate the main points in the project	Student does not address the topic well; minimal sharing about the project	Student is unprepared; no project available to share	___×6=___ (24 points)
Format	Creative, original expression of ideas in the best format	Conventional view, but aesthetically pleasing	Nothing new or original; surface-level treatment of the topic	Not enough information to make a judgment	___×4=___ (16 points)
Organization	Presentation includes a strong introduction that grabs audience attention, followed by sequentially organized progression of ideas	Presentation includes an introduction that gains audience attention followed by mostly organized progression of ideas	Presentation has a weak introduction and ideas are difficult to follow	The presentation is so disorganized that one cannot understand the language	___×3=___ (12 points)
Language	The language is vivid, imaginative with correct and colorful word choice and description	Efficient use of language, but needs variety of sentence length and style	Insufficient variety of sentences; lacks sophistication of language usage	Word choices are limited; there are many errors on language usage	___×3=___ (12 points)

Visual or Artifact Effects	Student uses visual/artifacts in unique and creative ways to enhance the message	Student uses effective visuals/artifacts to support the message	Student uses few visuals or artifacts to support the message	No visuals or artifacts are used in the presentation	___×5=___ (20 points)
Eye Contact	Student establishes good eye contact with audience and draws them into the presentation	During most of the presentation, student maintains eye contact with the audience	Student establishes minimal eye contact with the audience	Student establishes no eye contact with the audience	___×2=___ (8 points)
Vocal Projection	The speaker enunciates clearly, uses expression, and is clearly heard by the audience	The speaker enunciates clearly, but uses little expression	The pronunciation and enunciation are unclear; the volume is too low and/or the rate of speaking is too fast	The speaker does not enunciate and is very difficult to hear and understand	___×2=__ (8 points)
Body Language	The speaker uses gestures and mannerisms to enhance the delivery and does so naturally and effectively	The speaker uses some gestures and mannerisms to enhance the delivery	The speaker's mannerisms detract from the delivery	The speaker uses inappropriate mannerisms which distract the audience	___×1=___ (8 points)

Total Points (100 possible)

Grade Scale
A =
B =
C =
D =

Assessment Rubric for Art Exhibition

	Quality (4)	Acceptable (3)	Not Yet (2–1)
Effective Use of Elements of Art			
♦ Used interesting lines to create imagery	____	____	____
♦ Used creative shapes to create imagery	____	____	____
♦ Used effective colors to create imagery	____	____	____
♦ Used focal/vanishing points to create art	____	____	____
♦ Used a variety of textures (rough, smooth, shiny)	____	____	____
♦ Used empty space effectively	____	____	____
Effective Use of Principles of Art			
♦ Used balance to create a desirable effect	____	____	____
♦ Used variety and contrast to create an impression	____	____	____
♦ Used repetition (rhythm) to create an effect	____	____	____
♦ Used emphasis to make a point	____	____	____
♦ Used movement	____	____	____
♦ Used unity (harmony) to create an impression	____	____	____
Content			
♦ Project completed as assigned	____	____	____
♦ All criteria of assignment met	____	____	____
♦ Elements and principles blended well to create a message	____	____	____
♦ Used care in producing product	____	____	____
Overall Grade for the Product	____	____	____

Bibliography

Lee, H. (1999). *To kill a mockingbird.* New York: HarperCollins.

Part IV

Museum Assessments

16

THE LIVING LITERATURE MUSEUM

Ruth McClain and Margie Bush

To be successful at anything, the success must come gently—
with a great deal of effort, but with no stress or obsession.

Carlos Castaneda

Overview of Authentic Learning

A living literature museum involves students in a variety of research and performance activities. Students research an author or literary character from a national or thematic literature such as British, American, or Young Adult literature, depending on the focus of the class. Students complete a Biography Board to present their research on a person or event. Then they create costumes, design sets, and text that portray an accurate characterization of the individual chosen, assume the persona of that author or character as she/he would appear in a living museum, such as Williamsburg, Jamestown, or Sturbridge Village, and interact with the visitors/audience in either a frozen moment in time (as would be portrayed in a painting) or real-time scene (often called a "skit"), using, of course, the vernacular of the author/character and time period. This re-enactment activity may be lengthened or shortened to fit the time and ability levels of the class. This assessment is adapted from Ideas Plus, Book Seven (NCTE, 1989). For reenactments, see the C-SPAN American Writers Series.

Applications of this Activity

This activity is applicable to whole works (novels, plays, poems) as well as the usual excerpts appearing in survey anthologies. In fact, a bonus to this activity for classes that don't engage in supplemental readings of many full-length works is a more in-depth study of an author or character that is often prevented by the anthology's short excerpts.

What follows, then, is how this assessment might work in a typical high school literature class of American literature. For this discussion, the focus will be primarily on the study of the early to middle Romantic Period and Transcendentalism. There are several ways to implement the activity-two follow. In one implementation, the teacher distributes the Introduction to the assignment (see *Introduction to Students* on page 167) and each student adopts the persona of an author or character (see *Suggestions for Individual or Team Portrayals* on page 168), conducts research, using traditional and electronic sources, into the person and time period, designs a set, and acts out a living tableau of a pertinent moment in the author's or character's life; or the student poses as if in a picture frame, freezing that moment or scene. A second implementation of the activity asks students to join in teams to act out one or two moments from the life of the author/character or act out a pertinent scene from the selected story, poem, or novel (see *Requirements for Living Museum* on page 169). Regardless of implementation, students should be expected to interact with the visitors (classmates, teachers, parents) to the museum and engage in meaningful dialogue in the vernacular of the author/character and time period.

To broaden the scope of the activity, teachers might involve the school's art department, music department, and history department. Students could team up in each other's tableaux to personify artists, musicians, and other famous people who lived during the same time period as the chosen author or character.

NCTE/IRA Standards

Students will be expected to:

- compose texts that exhibit the conventions of mechanics, grammar, usage, and spelling appropriate to the audience, purpose, and topic (NCTE/IRA Standards #4, #5 and #6).
- research a character or author by gathering, evaluating, and synthesizing data from a variety of sources to communicate their discoveries in ways that suit their purpose and audience (NCTE/IRA Standard #7).
- use visual media and artifacts to communicate effectively with a variety of audiences and for a variety of purposes (NCTE/IRA Standard #8).
- role-play an author or character in an interactive living museum participating as knowledgeable, reflective, creative, and critical members of a variety of literacy communities (NCTE/IRA Standard #11), and use spoken, written, and visual language to accomplish their own purposes (NCTE/IRA Standard #12).

Performance Tasks

To show students what the Living Literature Museum might include and to review some of its characteristics, the class might begin by viewing the virtual tour of Plimouth Plantation (http://www.plimoth.org/). If possible, the class could also view a reenactment by professional actors of a famous author from C-SPAN's American Writers Series (http://www.c-spanstore.com/american-writers.html), which can be purchased from C-SPAN.

- The teacher distributes the Living Literature Museum Introduction, which contains the Time Line. (See *Introduction to Students.)*
- Individual Portrayals: Students choose an author or literary character that they admire and write their name on a sign-up sheet so that no two students portray the same person. (See *Suggestions for Individual or Team Portrayals* on page 168.)

Introduction to Students

Participants of the Living Literature Museum:

In this enrichment activity, the class will take on aspects of a museum, actually a "living museum," such as you might visit at Williamsburg, Sturbridge Village, or Plimoth Plantation. You will select, from the time period we are studying, an author or character to personify either by role-playing the person or by teaming with other students to reenact a moment from the author's life or a scene from a particular short story, poem, or novel. The Living Museum you design will be an interactive one. You are expected to dress, move, speak, pose, and engage knowledgeably in conversation with the museum's visitors as if you are the author or character. You will also prepare a brochure or flyer announcing your reenactment of the author or character; and write an invitation to other classes (to be determined later), teachers, and parents who might be interested. All of this will, of course, take considerable research using traditional book sources as well as electronic ones. As always, keep track of your sources so that they may be cited properly on your introductory flyer and on any other printed material you generate in this activity.

Please complete the following Planning Sheet (see page 171) and submit it after you have begun your research into the author or character. This Planning Sheet is due on _______________.

Suggestions for Individual or Team Portrayals

Early American Authors	Irving, Cooper, Emerson, Thoreau, Hawthorne, Melville, Twain, Crane, Cather
Early American Poets	Bryant, Poe, Longfellow, Dickinson, Whitman, Robinson, Masters
Modern Short Story Writers	Anderson, Hemingway, Faulkner, Fitzgerald, Porter, Steinbeck, Benét, Welty
Modern Poetry	Frost, Sandburg, Millay, Cummings, Stevens, Williams, Moore, Ranson, Macleish, Cullen, Auden, Roethke, Brooks, Lowell, Wilbur
Modern Drama	Wilder, O'Neill
Colonial Artists	Ralph Earl, John Copley, Charles Peale, Benjamin West, John Trumbull, Gilbert Stuart
Mystics and Realists	Albert Ryder, Ralph Albert Blakelock, Thomas Anshutz, Thomas Eakins, William Harnett, Winslow Homer
Modern American Artists	Grant Wood, Reginald Marsh, Jack Levine, Thomas Benton, Ben Shahn, Ivan Albright
Notable Characters	Tom Walker, Natty Bumppo, Ichabod Crane, Roderick Usher, Dr. Heidegger, Ishmael, Hiawatha, Hester Prinn, Pearl, Dimmesdale, Richard Cory, Neighbor Rosicky, Rip Van Winkle

- Team Portrayals: Students select a story, poem, or novel of interest to reenact and write their name on a sign-up sheet so that no two groups portray the same selection.
- Each student or team of students completes an *Activity Planning Sheet* (see on page 171) explaining what the Living Tableau will be, why that choice was made, how the choice fits into the unit of study, what roles each individual will play, and what peripherals will accompany the Living Tableau: music, set design or picture frame, artifacts, and so on. It might even ask the students to predict what sources they think they will need to consult.
- Students next use the various sources to research the person whom they have chosen. Students complete a Biography Board as evidence of the information they find. See *Biography Board* adapted from Stilley's *SCORE Teacher CyberGuide* (1997) and *Sample Biography Board* (on page 170).

Requirements for Living Museum

Due Date		*Assignment*
______________	1.	Select individual or team portrayal
______________	2.	Choose author or character and moment or scene
______________	3.	Begin initial research
______________	4.	Complete Planning Activity Sheet
______________	5.	Finish research
______________	6.	Prepare Biography Board
______________	7.	Prepare reenactment script and bibliography
______________	8.	Rehearsal for Living Literature Museum reenactment (all costumes, music, artifacts, etc., must be in the room and ready to go)
______________	9.	Complete the Self-Assessment Sheet
______________	10.	Schedule actual reenactment

Biography Board

- Title: In large letters, create a title using the name of the person or event.
- Focus Question: Write a focus question about your project.
- Illustration: Use photos or drawings of the individual and/or events.
- Poem: Write diamante, concrete, bio-poem, or I am poem.
- Time Line: Communicate important events in the person's life and how they relate to other historical events.
- Written Information: Should stress important details or contributions of the person.
- Quotes: Find significant quotes about the person or event.
- Captions: Type appropriate captions for your pictures.
- Questions: On the bottom of the backside of the biography board, write five questions that a student should be able to answer after reading your board.
- Both Sides: Use both sides to illustrate your individual or event.
- Name: In lower right-hand corner.

Sample Biography Board

- Students then choose one event from the person's life to portray or depict. They should keep the scene simple and consider what prompts will be necessary to accurately portray the setting in which they will place their character. The students decide how the character will stand, pose, dress, move, and talk. Students should also consider appropriate artifacts to include in the set. See *Activity Planning Sheet* on page 171.
- Students write invitations to other classes that might visit the museum, design flyers or posters announcing the museum, and schedule the time for the literary museum visitations.
- The teacher conducts a dress rehearsal and schedules breaks so that students do not tire. The teacher also assists students in setting up lighting and selecting music appropriate to the performance. See *Assessment Rubric for the Living Literature Museum* on page 173.

After the project is completed, each student or team reflects on the activity and fills out a *Self-Assessment* (see page 172), recording thoughts on such things as the actual research strategies used; successes and failures; audience reaction, etc.

Activity Planning Sheet

- Portrait of author/character:______________________________
____
____
- Reenacted by: ____
____
____
- Which part of the author's life, or to which moment in the character's life, do you want to reenact in a Living Literature Museum?
- Why do you think that this moment is of particular interest for you and/or for your audience?
- How does your choice fit into our unit of study?
- What roles will you (or you and your teammates) play?
- What peripherals will accompany the Living Tableau: consider the costumes, music, set design or picture frame, artifacts, etc.?
- What sources have you already consulted in reaching your decision?

Self-Assessment

Instructions: Once you have completed researching the author/character and the time period; have read the supplemental literature; and have prepared your costumes, music, set design, spoken script, etc., you will be ready to complete the following questionnaire.

- Rate Your Sources: 3 = good,
  2 = fair,
  1 = poor

Sources	*3*	*2*	*1*	*Comments*
Class textbook				
Other works by the same author				
Other literature anthologies				
Encyclopedias				
Biographies				
History texts				
Electronic sources—please list the best				
Other				

- Explain the failures, i.e., what were you unable to accomplish and why?
- What do you want your audience to know, feel, and think when you have completed the performance?
- What do you plan to do in your performance?
- Explain your successes. Of what are you most proud on this, the first part of the activity?

Assessment Rubric for the Living Literature Museum

	Distinguished (4)	*Proficient (3)*	*Apprentice (2)*	*Novice (1)*
Strength of Research Poster Organization and Appearance	Focused-purpose and subjects are defined clearly. Character is narrowed, researched, and organized. Audience is informed and enlightened.	Focus has some success; information is generally consistent. Some research evident; organization is generally clear. Audience is generally well-informed but some details need refinement	Focus is too broad or too narrow. Character is insufficiently researched or haphazardly delivered. Audience needs further information for clearer understanding.	Focus is not clearly defined—muddied. Character is very weak and insufficiently researched-lacks relevant details. Audience is left with unclear ideas or no ideas at all about the presentation.
Role-Play	Relaxed, self-confident, poised, dressed appropriately. Body language is natural, mood created helps audience visualize. Presentation aids are clear, appropriate, and beneficial to understanding.	Some nervousness exhibited; movements detract from appearance. Body language detracts from smooth presentation. Presentation aids add some clarity and definition to the character.	Unprepared, sloppy presentation of character, somewhat inadequately dressed. Body language is awkward; fidgeting, unnecessary movements. Presentation is somewhat inappropriate to character.	Totally inappropriate dress—out of character. Body language totally lacking. Presentation is totally inappropriate to character.
Visual Effects/Atmosphere of the Set	Visual appeal compels the viewer into the set. Motivates the viewer to linger, examine, and interpret the set. Excellent use of color, aesthetics, music, and artifacts.	The visual appeal is attractive and the viewer spends time looking at the set. The viewer is motivated to ask questions. Good atmosphere and authentic artifacts.	The visual appeal is somewhat lacking. The viewer takes a glance and moves on. Not much atmosphere created. Inappropriate artifacts or not typical for the character.	The set lacks character and the visual effects are inappropriate or hastily compiled. The viewer is nonplussed. There are no artifacts present.

Bibliography

National Cable Satellite Corporation (2001). *C-SPAN American Writers Series* [Online]. Available: http://www.americanwriters. org.

National Council of Teachers of English (1989). *Ideas plus, book seven.* Urbana, IL: NCTE.

Stilley, L. (1997). Biography board. In *SCORE: Teacher CyberGuide: Native American poetry.* San Diego: San Diego County Office of Education.

17

TOLERANCE MUSEUM AND CHILDREN OF THE HOLOCAUST DINNER PARTY

Jacqueline Glasgow

I believe in the sun even when it is not shining. I believe in love when feeling it not. I believe in God even when He is silent.

Inspiration on the walls of a cellar in Cologne, Germany, where Jews hid from the Nazis.

Overview of Authentic Learning

Prejudice, hatred, and anti-Semitism cannot go unnoticed or unchallenged. In this assessment, students will confront prejudice, moral choice and personal responsibility through research and role-play of a child who survived the Holocaust. They will create visual art, creative writing, and artifacts for a museum and dinner party. In order to get to know their character, students will create a Survivor Scrapbook and create monuments to commemorate a person, place, or event from their books. Students will plan the Survivor Dinner menu and the guest list. Each table of guests will include three children from the books on the booklist and an adult survivor. Students will script the conversation that will take place at the table, including feelings of the people, historical events, and historical facts. For the dinner party, students will cook the culturally appropriate foods and participate in the dinner according to their various roles.

NCTE/IRA Standards

During this assessment, students will:

- read Holocaust literature that explores the lives of Holocaust survivors. In doing so, they will read a wide range of print and nonprint texts to build an understanding of texts, of themselves, and of the cultures of the U.S. and the world; to acquire new information; to respond to the needs and demands of society and the workplace; and for personal fulfillment (NCTE/IRA Standard #1), and they will read a wide range of literature from many periods in many genres to build an understanding of the many dimensions of human experience (NCTE/IRA Standard #2).
- choose a particular survivor and create a character scrapbook. In doing so, students will apply a wide range of strategies to comprehend, interpret, evaluate, and appreciate texts (NCTE/IRA Standard #3); they will adjust their use of spoken, written, and visual language to communicate effectively with a variety of audiences and for different purposes (NCTE/IRA Standard #4); they will employ a wide range of strategies as they write and use different writing process elements appropriately to communicate with different audiences for a variety of purposes (NCTE/IRA Standard #5); and, they will apply knowledge of language structure, language conventions, media techniques, figurative language, and genre to create, critique, and discuss print and nonprint texts (NCTE/IRA Standard #6).
- create a monument to commemorate a person, place, or event in their books for the museum. In doing so, students will conduct research on issues and interests by generating ideas and questions, and by posing problems. They will gather, evaluate, and synthesize data from a variety of sources to communicate their discoveries in ways that suit their purpose and audience (NCTE/IRA Standard #7); and students will develop an understanding of and respect for diversity in language use, patterns, and dialects across cultures, ethnic groups, geographic regions, and social roles (NCTE/IRA Standard #9); and they will use spoken, written, and visual language to accomplish their own purposes (NCTE/IRA Standard #12).
- role-play a survivor of the Holocaust at a dinner party. In doing so, students will participate as knowledgeable, reflective, creative, and critical members of a variety of literacy communities (NCTE/IRA Standard #11); they will conduct research on issues and interests by generating ideas and questions, and by posing problems. They gather, evaluate, and synthesize data from a variety of sources to communicate their discoveries in ways that suit their purpose and audience (NCTE/IRA Standard #7); and they will adjust their use of spoken, written, and visual language to communicate effectively

with a variety of audiences and for different purposes (NCTE/IRA Standard #4).

Performance Tasks

- Students will read a variety of books from the booklist of children who survived the Holocaust. They might also visit the Museum of Tolerance at the Simon Wiesenthal Center to find photos and biographies of children caught up in the Holocaust at: http:// wiesenthal.com/mot/children/list1.cfm. Students of Jewish descent should research their own family backgrounds and experiences. They will then sign up for the child they would like to create a scrapbook for and then role-play at the Survivors' Dinner. See *Children Who Survived the Holocaust* for a sample listing of children (in literature) who survived the Holocaust.

Children Who Survived the Holocaust

Character	Book
Morris Kaplan	Adler's *One Yellow Daffodil*
Anita Lobel	Lobel's *No Pretty Pictures: A Child of War*
Elli Friedmann	Britton-Jackson's *I Have Lived a Thousand Years: Growing Up in the Holocaust*
Susie Weksler	Rabinovici's *Thanks to My Mother*
Ellen Rosen	Lowry's *Number the Stars*
Annmarie Johansen	Lowry's *Number the Stars*
Hannah	Yolen's *The Devil's Arithmetic*
Edith Knoll	Greenfeld's *The Hidden Children*
Gisele Warshawsky	Greenfeld's *The Hidden Children*
Frank Siegel	Greenfeld's *The Hidden Children*
Rosette Goldstein	Greenfeld's *The Hidden Children*
Alicja Melcer	Greenfeld's *The Hidden Children*
Roald Hoffmann	Greenfeld's *The Hidden Children*
Alice Sondike	Greenfeld's *The Hidden Children*
Paulette Pomeranz	Rosenberg's *Hiding to Survive*
Kurt Dattner	Rosenberg's *Hiding to Survive*
Rose Silberberg-Skier	Rosenberg's *Hiding to Survive*
Manny Stern	Rosenberg's *Hiding to Survive*
Cecile Rojer Jeruchim	Rosenberg's *Hiding to Survive*
Jacques Van Dam	Rosenberg's *Hiding to Survive*
Sylvia Richter	Rosenberg's *Hiding to Survive*
Andy Sterline	Rosenberg's *Hiding to Survive*

Hirsch Grunstein	Rosenberg's *Hiding to Survive*
Aviva Blumberg	Rosenberg's *Hiding to Survive*
Ruth Bachner	Rosenberg's *Hiding to Survive*
Debora Biron	Rosenberg's *Hiding to Survive*
Simon Jeruchim	Rosenberg's *Hiding to Survive*
Judith Steel	Rosenberg's *Hiding to Survive*
Corrie Ten Boom	*The Hiding Place*
Elie Wiesel	*Night*

- Students will create a Character Scrapbook for the child they have chosen to role-play. The scrapbooks can be constructed in photo albums, in PowerPoint Presentations, HyperStudio, or as Web pages. The information can be gleaned from books, Web sites, or the student's imagination. See *Survivor Scrapbook Requirements*.

Survivor Scrapbook Requirements

- Self-portrait or photo
- Found poem describing your character taken from your book
- Map of your country indicating the village, city, or farm where you live and showing the journey to the concentration camp
- Description or picture of your home and family
- Description of your school and friends
- Time line of historical events
- Storyboard of key events in your life
- Split-open head showing your life before and after the Holocaust
- Three journal entries describing your feelings and experiences
- Illustration of a favorite saying or quote
- Letter to a family member
- Thank you letters to people who helped you
- Picture and explanation of your hiding place or escape strategy
- Elegy or a lament of friends or family who died
- Story of your mental and physical threats, how your character manages to overcome these threats, and how your story ends
- Poems: *I Am, Bio,* or *Poem in Two Voices*
- Song lyrics to portray the emotions and concerns of your character along with your associations and connections
- Traditional favorite recipe

Paul Vogelman, a student of Jewish descent whose grandparents were survivors of the Holocaust, had the following to say about his writing:

> *Honestly, I feel like an imposter trying to portray the Holocaust in a piece of writing. I could see my grandparents forming a piece of writing from their feelings or personal experiences, but I feel out of place. The true survivors have the right to tell stories of the Holocaust. I can only reflect on how it appears to me. How can I, an 18-year-old of the late twentieth century even begin to scratch the surface of one of the most horrible events in history. Nevertheless, I do believe that creating this scrapbook and trying to carry on the memory of the Holocaust is better than not attempting to do so at all.*

- See *Assessment Rubric for Survivor Scrapbook* on page 180.
- As a final product, students create monuments to commemorate a person, place or event from their books. They should select a topic and then decide what medium would best reflect the objective of their memorial. Require students to write a descriptive summary that explains the reason for their choices and the symbolism represented. Students should plan to share their monuments and scrapbooks at the dinner. See *Jackie Horak's explanation of The World Must Remember* on page 181, taken from http:// chaucer.chesterfield.k12.va.us/Schools/Manchester_MS/hiding. html.

Assessment Rubric for Survivor Scrapbook

	Exemplary (4)	*Accomplished (3)*	*Developing (2)*	*Beginning (1)*	*Score*
Content	Covers assignments completely and in depth; insightful, creative, well-organized	Covers assignments well; informational, artistic; organized	Covers most of the assignments; somewhat interesting; mostly organized	Missing important assignments; not enough details; unorganized	__×10=__ (out of 40)
Journal Entries	Detailed, moving reflection on the thoughts, feelings, and associations made by the characters	Adequate reflection on the thoughts, feelings, and associations made by the characters	Insufficient reflection on the thoughts, feelings, and associations made by the characters	No evidence of reflection on the thoughts, feelings, and associations made by the characters	__×5=__ (out of 20)
Research	Research from more than three sources	Research evident from Internet and print material	Research from only one source	No research shown	__×5=__ (out of 20)
Poetry	Radiant images, sensuous, vivid use of language and poetic devices; unexpected comparisons are made	Clear attempt to create effective images using poetic language and devices; some use of metaphor and/or simile.	Limited command of poetic language and devices.	Denotative rather than connotative language; does not show command of basic literary devices	__×4=__ (out of 16)
Graphics/ Illustrations/ Maps	Exhibits mastery of skill in technique used to express creative ideas	Exhibits proficiency in technique used to express creative ideas	Exhibits some degree of skill in technique used to express creative ideas	Exhibits little or no apparent skill in technique used to express ideas	__×2=__ (out of 10)
Narration	Clear sense of time and place; well-developed plot; conflict resolution, change, discovery	Some sense of time and place; plot unfolds logically, but mechanically; reaches closure	Emerging sense of time and place; familiar plotline; skeletal outline; fails to reach closure	Little sense of time or place; no real storyline; details wander in search of plot; conflict or question not defined; random sequencing; list of events	__×4=__ (out of 20)
Letters	Letter content is clear, relevant, accurate, and concise	Letter content is mostly clear, relevant, accurate, and concise	Letter content is appropriate, but may lack a clear connection to the purpose	Little evidence of appropriate content	__×3=__ (out of 12)
Conventions	Correct grammar, punctuation, and spelling	Mostly correct grammar, punctuation, and spelling	Many errors in grammar, punctuation, and spelling	Numerous errors in grammar, punctuation, and spelling	__×2=__ (out of 8)
			Total Points (146 possible)		
			Grade Scale	A = B = C = D =	

The World Must Remember

The world must remember what monstrous acts mankind is capable of committing against others. Rising from the bones, to create the sensation of a mass grave of Holocaust victims, a Star of David symbolizes hope. Once worn by the victims as a means of identification, the star is symbolically supported by the bones. The bones surround the base of the Star of David, but they do not hold it. The Star of David rises toward the heavens to shine a warning, as well as a message of hope. An excerpt from the poem "In All Those Camps" hides itself within the marbled pattern of the star, becoming visible to the viewer only upon close inspection, just as many life's important messages are also hidden. Only the boldness of the monument can entice the viewer to study it long enough to find the hidden message. The marble is shiny and allows the viewer to see his reflection in the monument, and yet it is solid and heavy enough to make clear the seriousness of the message.

- Students then prepare for the Holocaust Survivor Dinner Party. They need to select the date, make invitations, determine the agenda, plan the food, select the music, and determine the entertainment and conversation. If the dinner is to be potluck, then students should prepare the recipe they have selected and presented for their character in the scrapbook. See *Assessment Rubric for Holocaust Museum and Survival Dinner* on page 182.

Assessment Rubric for Holocaust Museum and Survival Dinner

	Exemplary (4)	*Acceptable (3)*	*Limited (2)*	*Unacceptable (1)*
Monument	Creative artifact and explanation that represents some part of the survivor's Holocaust experience.	Acceptable artifact and explanation that represents the survivor's Holocaust experience.	Vague or incorrect representation and explanation of the survivor's Holocaust experiences.	No explanation of the survivors Holocaust experiences.
Participation in Conversation	Creative, informative script based on correct information, historical facts, events, and feelings of the character.	Acceptable research and communication of the character's biography, life, and feelings.	Limited research and communication of the character's biography, experiences, and feelings.	No evidence of research and communication of the character's biography, experiences, and feelings.
Body Language	Student uses gestures and mannerisms to enhance the delivery of the conversation, and does so naturally and effectively.	Student uses some gestures and mannerisms to enhance the delivery of the conversation.	Student uses some gestures and mannerisms that detract from the delivery of the conversation.	Student's behavior distracts the audience and is inappropriate.
Food Contribution	Authentic and culturally appropriate. Tasty; appetizing!	Culturally appropriate, but not unique.	Common, ordinary, not representative of the culture.	No contribution made.
Presentation of Artifact and Scrapbook	Student shares projects creatively and provides vivid examples to illustrate the character's experiences during the Holocaust.	Student shares projects clearly, but provides few examples of the character's life and experiences.	Student is unprepared to share his/her projects and tell stories about his/her character's survival.	No artifact to share.

Bibliography

Ackerman, K. (1994). *The night crossing.* New York: Knopf.

Adler, D. (1994). *Hilde and Eli: Children of the Holocaust.* New York: Holiday House.

Adler, D. (1987). *The number on my grandfather's arm.* New York, UAHC Press.

Adler, D. (1995). *One yellow daffodil.* New York: Gulliver Books.

Adler, D. (1989). *We remember the Holocaust.* New York: Holt.

Auerbacherk, I. (1986). *I am a star: Child of the Holocaust.* Puffin Books.

Britton-Jackson, L. (1997). *I have lived a thousand years: Growing up in the Holocaust.* New York: Simon & Schuster.

Brook, P. (1996). *The United States Holocaust Memorial Museum.* New York: Children's Press.

Fisch, R. (1994). *Light from the yellow star.* Minneapolis, MN: Yellow Star Foundation.

Frank, A. (1993). *The diary of a young girl.* New York: Simon & Schuster.

Greenfeld, H. (1993). *The hidden children.* New York: Houghton Mifflin.

Grossman, M. (2000). *My secret camera: Life in the Lodz ghetto.* New York: Gulliver Books.

Handler, A., & Meschel, S. V. (1993). *Young people speak: Surviving the Holocaust in Hungary.* New York: Franklin Watts.

Hoestlandt, J. (1993). *Star of fear, star of hope.* New York: Walker and Company.

Innocenti, R. (1985). *Rose Blanche.* Creative Education.

Klein, G. W. (1995). *All but my life.* New York: Hill and Wang.

Komski, J. (2001). *The art of Auschwitz survivor* [Online]. Available: http://remember.org/ komski/index.html.

Laird, C. (1990). *Shadow of the wall.* New York: Green Willow.

Leapman, M. (2000). *Witnesses to war.* New York: Scholastic.

Lobel, A. (1998). *No pretty pictures: A child of war.* New York: Greenwillow Books.

Lowry, L. (1989). *Number the stars.* New York, Bantam Doubleday.

Marks, J. (1993). *The hidden children: The secret survivors of the Holocaust.* New York: Fawcett Columbine.

Matas, C. (1993). *Daniel's war.* New York: Scholastic.

Matas, C. (1989). *Lisa's war.* New York: MacMillan.

Rabinovici, S. (1998). *Thanks to my mother.* New York: Dial Books.

Richter, H. P. (1987). *Friedrich.* Puffin Books.

Rogasky, B. (1988). *Smoke and ashes: The story of the Holocaust.* New York: Holiday House.

Rosenberg, M. (1994). *Hiding to survive.* New York, Clarion.

Saldinger, A. (2001). *Life in a Nazi concentration camp.* San Diego, Lucent Books.

Spiegelman, A. (1973). *Maus: A survivor's tale I: My father bleeds history.* New York: Pantheon Books.

Spiegelman, A. (1986). *Maus: A survivor's tale II: And there my troubles began.* New York: Pantheon Books.

Stewart, G. B. (1995). *Life in the Warsaw ghetto.* San Diego: Lucent Books.

Ten Boom, C. (1999). *The hiding place.* New York: Econo-Clad Books.

Volavkova, H. (Ed.). (1993). *I never saw another butterfly: Children's drawings and poems from Terezin concentration camp 1942–1944.* New York: Schocken Books.

Wiesel, E. (1982). *Night.* New York: Bantam Doubleday Dell Publishing.

Yolen, J. (1988). *The devil's arithmetic.* New York: Puffin Books.

REFERENCES

(1997). *Ellis Island* [Video]. A & E Entertainment.

(2000, April). *Jekyll and Hyde* [Online]. Available: http://jekyll-hyde.com/welcome-b.html.

Ackerman, K. (1994). *The night crossing.* New York: Knopf.

Adler, D. (1987). *The number on my grandfather's arm.* New York, UAHC Press.

Adler, D. (1989). *We remember the Holocaust.* New York: Holt.

Adler, D. (1994). *Hilde and Eli: Children of the Holocaust.* New York: Holiday House.

Adler, D. (1995). *One yellow daffodil.* New York: Gulliver Books.

Adoff, A. (1995). *Slow dance heart break blues.* New York: Lothrop, Lee & Shepard Books.

Angelou, M. (1996). *I know why the caged bird sings.* New York: Random House.

Anonymous (1994). *Go ask Alice.* New York: Simon & Schuster.

Auerbacherk, I. (1986). *I am a star: Child of the Holocaust.* Puffin Books.

Baie Comeau High School (October 31, 2001). *Reader's theatre* [Online]. Available: http://www.qesn. meq.gouv.qc.ca/schools/bchs/rtheatre/teach2.htm.

Bloor, E. (1997). *Tangerine.* New York: Harcourt Brace.

Bloor, E. (1999). *Crusader.* New York: Harcourt Brace.

Bridgers, S. E. (1996). *All we know of heaven.* Wilmington, NC: Banks Channel Books.

Britton-Jackson, L. (1997). *I have lived a thousand years: Growing up in the Holocaust.* New York: Simon & Schuster.

Brook, P. (1996). *The United States Holocaust Memorial Museum.* New York: Children's Press.

Canady, R. L., & Rettig, M. D. (1996). *Teaching in the block: Strategies for engaging active learners.* Princeton, NJ: Eye On Education.

Carlson, L. M. (1994). *Cool salsa: Bilingual poems on growing up Latino in the United States.* New York: Henry Holt.

Carson, J. (1989). *Stories I ain't told nobody yet.* New York: Orchard Books.

Chicago Board of Education (2002). *Massachusetts speaking assessment criteria. Assessment of basic skills, speaking assessment rating guide* [Online]. Available: http://intranet.cps.k12.il.us/Assessments/ideas_and_rubrics/intro_scoring/intro_scorng.html.

Cormier, R. (1988). *Fade.* New York: Delacorte.

Cormier, R. (1991). *We all fall down.* New York: Delacorte.

Cormier, R. (1999). *In the middle of the night.* New York: Econo-Clad Books.

Cronyn, G. W. (Ed.). (1991). *American Indian poetry: An anthology of songs and chants.* New York: Ballantine Books.

Crutcher, C. (1987). *The crazy horse electric game.* New York: William & Morrow.

Crutcher, C. (1991). *Athletic shorts.* New York: Greenwillow Books.

Crutcher, C. (1996). *Chinese handcuffs.* New York: Dell Publishing.

C-SPAN American Writers Series [Online]. Available: http://www.americanwriters.org.

Cullinan, B. E. (Ed.). (1996). *A jar of tiny stars: Poems by NCTE award-winning poets.* New York: Boyds Mills Press.

Curtis, C. (1999). *Bud, not buddy.* New York: Delacorte Press.

DeSoto, G. (1992). *A fire in my hands.* New York: Scholastic.

Dunbar, P. L. (2000, April). *We wear the mask* [Online]. Available: http://www.library.utoronto.ca/utel/rp/poems/dunbarp5.html.

Dunning, S., Lueders, E., & Smith, H. (1967). *Reflections on a gift of watermelon pickle... and other modern verse.* New York: Lothrop, Lee & Shepard Books.

Dunning, S., & Stafford, W. (1992). *Getting the knack: 20 poetry writing exercises.* Urbana, IL: National Council of Teachers of English.

Edmiston, B. (1998). Drama as inquiry: Teachers and students as co-researchers. In J. Wilhelm & B. Edmiston, *Imagining to learn; inquiry, ethics, and integration through drama.* Portsmouth, NH: Heinemann.

"Façade." CD of Broadway musical *Jekyll and Hyde* [Online]. Original Broadway cast recording. Performers include Robert Cuccioli, Linda Eder, and Christiane Noll. (No date given.)

Falk, B. (2000). *The heart of the matter: Using standards and assessment to learn.* Portsmouth, NH: Heinemann.

Farell, E. (1996). In *Standards for the English language arts.* Urbana, IL: National Council of Teachers of English/International Reading Association.

Fisch, R. (1994). *Light from the yellow star.* Minneapolis, MN: Yellow Star Foundation.

Fitzgerald, F. S. (1996). *The great Gatsby.* New York: Scribner and Sons.

Fleischman, P. (1988). *Joyful noise: Poems for two voices.* New York: HarperTrophy.

Fleischman, P. (1989). *I am phoenix: Poems for two voices.* New York: HarperTrophy.

Fleischman, P. (1998) *Whirligig.* New York: Henry Holt.

Fleischman, P. (1999). *Seedfolks.* New York: HarperTrophy.

Fox, J. (1997). *Poetic medicine.* New York: Putnam.

Frank, A. (1993). *The diary of a young girl.* New York: Simon & Schuster.

Freedman, R. (1993). *Eleanor Roosevelt: A life of discovery.* New York: Clarion Books.

Garden, N. (1982). *Annie on my mind.* New York: Farrar, Straus, Giroux.

Gibaldi, J. (1999). *MLA handbook for writers of research papers* (5th ed.). New York: The Modern Language Association of America.

Glasgow, J., & Bush, M. (1996, May). Students use their multiple intelligences to develop promotional magazines for local businesses. *Journal of Adolescent and Adult Literacy,* 39, 638–649.

Glasgow, J., & Bush, M. (1995, December). *Promoting active learning and collaboration through a marketing project.* English Journal, 84, 32–37.

Glatthorn, A. A. (1999). *Performance standards authentic learning.* Larchmont, NY: Eye on Education.

Glenn, M. (1982). *Class dismissed! high school poems.* New York: Clarion.

Glenn, M. (1986). *Class dismissed II: More high school poems.* New York: Houghton Mifflin.

Glenn, M. (1988). *Back to class: Poems by Mel Glenn.* New York: Clarion.

Glenn, M. (1991). *My friend's got this problem, Mr. Candler.* New York: Clarion.

Glenn, M. (1996). *Who killed Mr. Chippendale?: A mystery in poems.* New York: Lodestar Books.

Glenn, M. (1997). *The taking of room 114: A hostage drama in poems.* New York: Lodestar Books.

Greene, B. (1974). *Summer of my German soldier.* New York: Bantam Press.

Greenfeld, H. (1993). *The hidden children.* New York: Houghton Mifflin.

Grossman, M. (2000). *My secret camera: Life in the Lodz ghetto.* New York: Gulliver Books.

Guterson, D. (1994). *Snow falling on cedars.* San Diego: Harcourt Brace.

Guthrie, W. (2000). Dust Bowl ballads audio CD, "Dust Bowl refugee." New York: BMG/Buddah Records.

Handler, A., & Meschel, S. V. (1993). *Young people speak: Surviving the Holocaust in Hungary.* New York: Franklin Watts.

Hawthorne, N. (1992). *The scarlet letter.* New York: Knopf.

Hesse, K. (1997). *Out of the dust.* New York, Scholastic.

Hoestlandt, J. (1993). *Star of fear, star of hope.* New York: Walker and Company.

Huxley, A. (1998). *Brave new world.* New York: HarperCollins Perennial Library.

Innocenti, R. (1985). *Rose Blanche.* Creative Education.

Interact Middle School Catalogue (2001). *Gateway: A simulation of American immigration history.* Carlsbad, CA: author (available: www.interact-simulations.com/).

Janeczko, P. B. (1984). *Poetry from A to Z: A guide for young writers.* New York: Bradbury Press.

Janeczko, P. B. (1984). Strings: A gathering of family poems. New York: Bradbury.

Janeczko, P. B. (1985). Pocket poems: Selected for a journey. New York: Bradbury Press.

Janeczko, P. B. (1990). *The place my words are looking for.* New York: Bradbury Press.

Janeczko, P. B. (1991). *Poetspeak: In their work, about their work.* New York: Collier Books.

Janeczko, P. B. (1991). *Preposterous: Poems of youth.* New York: Orchard Books.

Janeczko, P. B. (Ed.). (1993). *Looking for your name: A collection of contemporary poems.* New York: Orchard Books.

Kazemek, F. E., & Rigg, P. (1995). *Enriching our lives: Poetry lessons for adult literacy teachers and tutors.* Newark, DE: International Reading Association.

Kerr, M. E. (1986). *Deliver us from Evie.* New York: HarperCollins.

Kerr, M. E. (1997). *Hello, I lied.* New York: HarperCollins.

Klein, G. W. (1995). *All but my life.* New York: Hill and Wang.

Komski, J. (2001). *The art of Auschwitz survivor* [Online]. Available: http://remember.org/komski/index.html.

Laird, C. (1990). *Shadow of the wall.* New York: Green Willow.

Leapman, M. (2000). *Witnesses to war.* New York: Scholastic.

Lee, H. (1999). *To kill a mockingbird.* New York: HarperCollins.

Lobel, A. (1998). *No pretty pictures: A child of war.* New York: Greenwillow Books.

Lowry, L. (1989). *Number the stars.* New York, Bantam Doubleday.

Lowry, L. (1993). *Gathering blue.* New York: Houghton Mifflin.

Manna, A. (1992). *Young adult literature course matherials.* Kent, OH: Kent State University.

Marks, J. (1993). *The hidden children: The secret survivors of the Holocaust.* New York: Fawcett Columbine.

Marzano, R.J., & Kendall, J.S. (1996). *A comprehensive guide to designing standards-based districts, schools, and classrooms.* Alexandria, VA: Association for Supervision and Curriculum Development.

Matas, C. (1989). *Lisa's war.* New York: MacMillan.

Matas, C. (1993). *Daniel's war.* New York: Scholastic.

McCaslin, N. (1990). *Creative drama in the classroom* (5th ed.). White Plans, NY: Longman.

Mehlich, S., & Smith-Worthington, D. (1997). *Technical writing for success:* A school-to-work approach. Cincinnati, OH: South-Western Educational Publishing.

Myers, M., & Spalding, E. (1997). *Standards exemplar series: Assessing student performance grades 9–12.* Urbana, IL: National Council of Teachers of English.

Myers, W. D. (1988). *Fallen angels.* New York: Scholastic Books.

National Council of Teachers of English. (1989). Ideas plus, book seven. Urbana, IL: NCTE.

Nilsen, A. P., & Donelson, K. L. (2001). *Literature for today's young adults,* (6th ed.). New York: Longman.

O'Brien, R. C. (1974). *Z for Zachariah.* New York: Aladdin.

O'Brien, T. (1990). *The things they carried.* Boston: Houghton Mifflin.

O'Neill, C. (1995). *Drama worlds: A framework for process drama.* Portsmouth, NH: Heinemann.

Paulsen, G. (1997). *Sarny.* New York: Delacorte Press.

Porter, T. (1997). *Treasures in the dust.* New York: Harper/Trophy.

Prescott, J. (Ed.). (1995). *America at the cross roads: Great photographs from the thirties.* New York: Bromption Books.

Rabinovici, S. (1998). *Thanks to my mother.* New York: Dial Books.

Raven, M. T. (1997). *Angels in the dust.* Mahwah, NJ: Troll Communications.

Richter, H. P. (1987). *Friedrich.* Puffin Books.

Rogasky, B. (1988). *Smoke and ashes: The story of the Holocaust.* New York: Holiday House.

Romano, T. (1995). *Writing with passion.* Portsmouth, NH: Heinemann.

Romano, T. (2000). *Blending genre, altering style: Writing multigenre papers.* Portsmouth, NH: Boynton/Cook.

Rosenberg, M. (1994). *Hiding to survive.* New York, Clarion.

Rowling, J. K. (1998). *Harry Potter and the sorcerer's stone.* New York: Scholastic.

Rowling, J. K. (1999). *Harry Potter and the chamber of secrets.* New York: Scholastic.

Rowling, J. K. (1999). *Harry Potter and the prisoner of Azkaban.* New York: Scholastic.

Rowling, J. K. (2000). *Harry Potter and the goblet of fire.* New York: Scholastic.

Rubrics for Web lessons [Online]. Available: http://edweb.sdsu.edu/webquest/rubrics/welessons.htm.

Rylant, C. (1994). *Something permanent.* New York: Harcourt Brace.

Ryskamp, G., & Ryskamp, P. (1996). *A student's guide to Mexican American genealogy. Oryx American Family Tree Series.* Westport, CT: Oryx Press.

Sachar, H. (1999). *The boy who lost his face.* New York: Econo-Clad Books.

Saldinger, A. (2001). *Life in a Nazi concentration camp.* San Diego, Lucent Books.

SCORE: *Teacher guide-Night by Elie Weisel* [Online]. Available: http://www.sdcoe.d12.ca.us/score/night/nighttg.html.

Sebranek, P., Meyer, V., & Kemper, D. (1992). *Writers INC: A guide to writing, thinking, & learning.* Burlington, WI: Write Source Educational Publishing House.

Shepard, A. (1996). *Aaron Shepard's RT page* [Online]. Available: http://www.aaronshep.com/rt/wahtis.html.

Short, K., Harst, J., & Burke, C. (1996). *Creating classrooms for authors and inquirers.* Portsmouth, NH: Heinemann.

Spiegelman, A. (1973). *Maus: A survivor's tale I: My father bleeds history.* New York: Pantheon Books.

Spiegelman, A. (1986). *Maus: A survivor's tale II: And there my troubles began.* New York: Pantheon Books.

Spinelli, J. (1997). *Wringer.* New York: HarperCollins.

Standards for the English language arts (1996). Urbana, IL: National Council of Teachers of English/International Reading Association.

Stanley, J. (1992). *Children of the Dust Bowl.* New York: Crown.

Starr, L. (2000). *A good rubric.* Education World.

Stewart, G. B. (1995). *Life in the Warsaw ghetto.* San Diego: Lucent Books.

Stiggins, R. J. (1997). *Student-centered classroom assessment* (2nd ed.). Columbus, OH: Merrill.

Stilley, L. (1997). Biography board. In *SCORE: Teacher CyberGuide: Native American poetry.* San Diego: San Diego County Office of Education.

Ten Boom, C. (1999). *The hiding place.* New York: Econo-Clad Books.

Tolan, S. (1996). *Welcome to the ark.* New York: Morrow Junior Books.

Voigt, C. (1994). *When she hollers.* New York: Scholastic.

Volavkova, H. (Ed.). (1993). *I never saw another butterfly: Children's drawings and poems from Terezin concentration camps 1942–1944.* New York: Pantheon Books.

Walker, L. (2001). *Readers' theatre reading resources and teaching tools children will love* [Online]. Available: http://loiswalker.com/catalog/teach.html.

Weisel, E. (1982). *Night.* New York: Bantam Books.

Wolsch, R. A., & Wolsch, L. A.C. (1988). *From speaking to writing to reading: Relating the arts of communication.* New York: Columbia University Teachers College.

Yolen, J. (1988). *The devil's arithmetic.* New York: Puffin Books.